SpringerBriefs in Optimization

SpringerBriefs in Optimization showcases algorithmic and theoretical techniques, case studies, and applications within the broad-based field of optimization. Manuscripts related to the ever-growing applications of optimization in applied mathematics, engineering, medicine, economics, and other applied sciences are encouraged.

More information about this series at http://www.springer.com/series/8918

Demetrios Serakos

Generalized Adjoint Systems

Springer

Demetrios Serakos
Dahlgren, VA, USA

ISSN 2190-8354 ISSN 2191-575X (electronic)
SpringerBriefs in Optimization
ISBN 978-3-319-16651-3 ISBN 978-3-319-16652-0 (eBook)
DOI 10.1007/978-3-319-16652-0

Library of Congress Control Number: 2015935202

Mathematics Subject Classification (2010): 93A30, 47A05, 93A10, 46B25

Springer Cham Heidelberg New York Dordrecht London

Printed on acid-free paper

Springer International Publishing AG Switzerland is part of Springer Science+Business Media (www.springer.com)

ΜΝΗΜΗ ΓΕΩΡΓΙΑ (ΠΑΝΤΕΛΟΠΟΥΛΟΣ) ΣΥΡΡΑΚΟΣ ΚΑΙ
ΔΗΜΗΤΡΙΟΣ ΣΥΡΡΑΚΟΣ

Foreword

This book is the result of a theoretical development and examination of the generalized adjoint concept and the conditions under which systems analysis using adjoints is valid. The results developed in this book are useful aids for the analysis and modeling of physical systems, including the development of guidance and control algorithms, and in developing simulations.[1] This book defines and develops the generalized adjoint of an input–output system. The generalized adjoint of a system may be used to represent the inverse of a system, which may be used in systems analysis. Furthermore, the generalized adjoint condenses system behavior in a way that facilitates analysis of the effect of noise and disturbance on a system. An earlier version of this book appeared as a Naval Surface Warfare Center, Dahlgren Division technical report.[2]

The author is indebted to Professor W.L. Root (1919–2007) for his help with the earlier drafts which became this book. Also, thanks to Ms. Razia Amzad from Springer.

[1]John E. Gray.

[2]D. Serakos, "Generalized Adjoint Systems," NSWCDD/TR12/79, Naval Surface Warfare Center, Dahlgren Division, Dahlgren, VA 22448. Posted on www.dtic.mil. This version was not sponsored by NSWCDD and does not necessarily reflect the views or policies of the U.S. Navy or U.S. Government.

Contents

Contents

Nomenclature

u, v	Input, Page 8
$u_{s,t}, v_{s,t}$	Truncated input, Definition 1 and Page 6, 8
u_t, v_t, u^t, v^t	Truncated (past) input, Page 8, 9
$u_{t,\infty}, v_{t,\infty}$	Future input, Page 8
H	Bounding space containing inputs, Page 7
K	Bounding space containing outputs, Page 9
U	Input space, Page 9
U_t	Truncated (past) input space, Page 9
$U_{t,\infty}$	Future input space, Page 9
Y, Y_t	Output space, truncated (past) output space, Page 9
F	System map, Definition 4, Page 9
$\mathcal{F}$	Space of system maps, Page 21
$\tilde{F}_t$	Truncated system map, Page 9
F_t	Centered truncated system map, Page 9
(Y, F, U)	Input–output system, Definition 4, Page 9
Y^a, U^a	Generalized adjoint input and output spaces, Definition 10, Page 18
$F^a, (U^a, F^a, Y^a)$	Generalized adjoint system, Definition 10, Page 18
$\mathcal{F}^a$	Space of generalized adjoint systems, Page 21
$\mathrm{L}_t, \mathrm{R}_t$	Left-, right-shift by t, Definition 1, Page 9
$\alpha_U, \mathcal{K}_U$	α, $\mathcal{K}$ from (5) of Definition 1, associated with U, Page 6
$\mathcal{D}_N(U, Y)$	Bounded operators U to Y, Page 11
$\mathcal{C}_N(U, Y)$	Bounded, continuous operators U to Y, Page 11
$\lVert\cdot\rVert$	Vector norm, Page 6
$\lVert\cdot\rVert_{s,t}$	Fitted family seminorm over (s, t), Definition 1, Page 6
$[\![\cdot]\!]$	Operator norm, Eq. (2.6), Page 10

$\nrightarrow$	Splice, Definition 3, Page 8
$\phi : \mathcal{F} \rightarrow \mathcal{F}^a$	The generalized adjoint map, Chapter 5, Page 21
$A^\#, B^\#$	Auxiliary input and output systems, Equation (6.5), Page 34
$\gamma : U \rightarrow \mathcal{A}^\#$	Map from input space to auxiliary input space, Page 30
$G, \mathcal{G}$	Auxiliary map; space of auxiliary maps (6.22), Page 35
$\rho : \mathcal{F} \rightarrow \mathcal{G}$	System to auxiliary system map, Page 37
$\square$	Indicates the end of a proof, Page 15, 19, 22–24, 26, 28–34, 37, 38, 40, 42, 59–61

Chapter 1
Introduction

Generalized adjoint systems are defined and analyzed in this book. A generalized adjoint system is an input–output system that is derived from a given causal (possibly nonlinear) input–output system. The basic properties of the generalized adjoint system are that it is linear, causal (in fact, memoryless), and is time invariant when the original system is time invariant. Given a space of causal input–output systems, a map from this space of systems to its generalized adjoints, referred to as the adjoint map, is defined. The adjoint map is bounded, continuous, and preserves translations. Under certain conditions, and involving the Hahn–Banach theorem, the adjoint map has a bounded and continuous inverse. A representation for the inverse of a causal input–output system in terms of the generalized adjoint is given. It is seen that if the original input–output system is one to one/onto, then its generalized adjoint is onto/one to one. The reverse of these implications may be obtained under some completeness conditions. In the first step of developing a representation of the inverse to the original input–output system using the generalized adjoint system, another input–output system, which has the behavior of the original input–output system, is defined. This input–output system is referred to as the auxiliary input–output system. Conditions for the auxiliary input–output system to be causal, bounded, and continuous are given. Conditions for the map between the original input–output system and the auxiliary input–output system to be invariant with respect to translations, bounded, continuous, and to have a continuous inverse are given. It is then shown that the generalized adjoint system is a representation for the inverse of the auxiliary input–output system and therefore also of the original input–output system. A method for computing bounds for undesired inputs to an input–output system involving the generalized adjoint system is discussed. An example illustrating Chaps. 4, 5, and some of 6 is presented. Portions of this book appeared in [14, 16, 19].

Inverses are used in systems theory, see Kailath [7], for example. Adjoint methods are used in the analysis of systems. Generally, adjoint methods are used in simulations because a lot of information can be obtained with a low number

D. Serakos, *Generalized Adjoint Systems*, SpringerBriefs in Optimization,
DOI 10.1007/978-3-319-16652-0_1

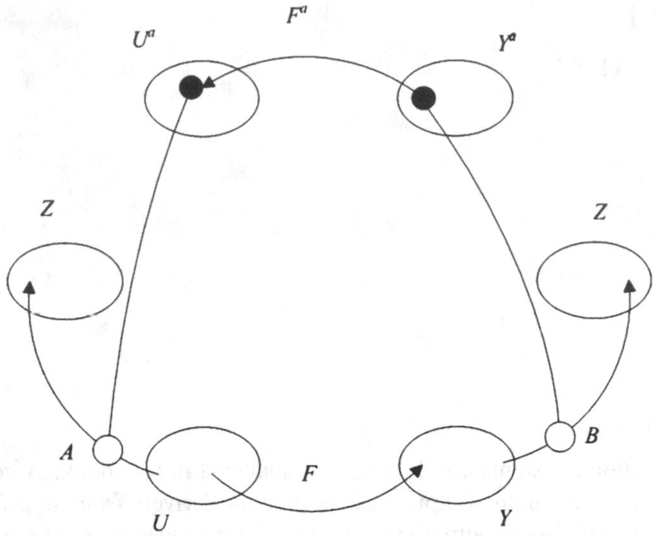

Fig. 1.1 The generalized adjoint system

of adjoint runs. We give three examples. Campobasso et al. use adjoints to help design turbomachinery blades [1]. Pires and Miranda use adjoints in the analysis of tsunami [10]. Morency et al. use adjoints to study wave propagation [9]. The effects of undesired inputs, e.g., noise and disturbances, in systems theory is important. For example, see Freudenberg et al. [5], and Serakos [18].

A rough description of the generalized adjoint system is illustrated in Fig. 1.1 A precise description is given in Chap. 3. It is best to keep in mind the construction of the adjoint of a linear transformation in Banach space. Let $\mathcal{B} = \{(Z, B, Y)\}$ be a set of causal input–output systems, where Y is the input space of time functions and Z is the output space of time functions and $B : Y \to Z$ is the system map. For $B \in \mathcal{B}$, a truncated input–output system $B_t : Y_0 \to Z_0$ is defined for each $t \in \Re$, where Y_0 (Z_0) is the input (output) space of truncated time functions defined up to time zero. In this scenario, the system trajectory $t \to B_t$ may be considered as a time function. This type of set, namely $\mathcal{B}$, will be the input space of a generalized adjoint system. Now, consider a causal input–output system (Y, F, U). A truncated system $F_t : U_0 \to Y_0$ may be defined for each $t \in \Re$. By concatenating a $B \in \mathcal{B}$ with F, another causal input–output system is defined: $A = BF$. In fact, a set of causal input–output systems $\mathcal{A} = \{(Z, A, U)\}$ is defined by concatenating every $B \in \mathcal{B}$ with F. The system trajectories $t \to A_t$ of $\mathcal{A}$ may also be considered as time functions. A *generalized adjoint system* is defined as the mapping from the set of time functions $\mathcal{B}$ to the set of time functions $\mathcal{A}$.

Denote an input space of time functions $Y^a = \{(Z, B, Y)\}$ and an output space of time functions $U^a = \{(Z, A, U)\}$. Given an input–output system (Y, F, U), its generalized adjoint is $F^a : Y^a \to U^a$ defined by $F^a(B) = A = BF$. The generalized adjoint system is (U^a, F^a, Y^a). Consider a set of causal input–output systems $\mathcal{F} = \{(Y, F, U)\}$. A *generalized adjoint map* with domain $\mathcal{F}$ and range $\mathcal{F}^a$ is defined by $\phi(F) = F^a$ for all $F \in \mathcal{F}$.

The results presented in this book are original and the author's, except as stated in the foreword.

Chapter 2
Preliminaries

This chapter presents a workable definition of an input–output system. An input–output system is herein denoted (Y, F, U) where F is a mapping from an input space U to an output space Y, and where U and Y are translation-invariant spaces of vector-valued time functions. The vector values of these time functions typically are $\Re^N$, $N = 1, 2, \ldots$, but only need to be Banach spaces. Other spaces and mappings related to (Y, F, U) will be introduced, but U and Y with the qualifying affixes they carry always refer to input and output spaces. Mappings from various input spaces to output spaces are denoted F, again with qualifying affixes.

Among the properties discussed are boundedness and continuity; therefore, a topology with metric on the sets ("spaces") involved, including the spaces of mappings must be provided.[1] For convenience and because it is appropriate to deal with linear spaces, or subsets thereof, we choose to do this by introducing norms (or semi-norms) on these linear spaces. Of course, the introduction of particular norms is arbitrary, but there are considerations to take into account. For instance, it is desirable: (1) to have generality where possible (i.e., to deal with classes of norms rather than specific ones), (2) to use, or permit the use of, norms that are readily calculable, and (3) to use, or permit the use of, norms that yield a satisfactory interpretation in physical problems (an admittedly vague statement). From these considerations we are led first to define classes of norms and normed linear spaces to be used in specifying input and output spaces for systems. These classes are referred to as Fitted Families (FFs) of normed linear spaces. Roughly speaking, fitted families work like L_p norms on time functions with the additional feature where a time weighting can be incorporated so the distant past of an input or output time function may be de-emphasized.[2] This feature is not as important in this book

[1] We don't need metrics for continuity, of course.

[2] Desoer and Vidyasagar discuss systems theory with L_p norms, [3].

D. Serakos, *Generalized Adjoint Systems*, SpringerBriefs in Optimization,
DOI 10.1007/978-3-319-16652-0_2

as it is in [17, 20] or [21]. The notation $\|u_{s,t}\|_{s,t}$ indicates the norm; e.g., a weighted L_p norm, of the input u over the interval of time $(s, t]$. $U_{s,t}$ is the space of inputs over the same interval. FFs were initially described in [11]. Additional statements motivating the use of fitted families are in [15].

Definition 1 ([11]). Let $\mathcal{L} = \mathcal{L}(\Re, E)$ be a linear space of time functions from $\Re$ into a Banach space E such that any translate of a function in $\mathcal{L}$ is also a function in $\mathcal{L}$. Let $\mathcal{N} = \{\|\cdot\|_{s,t}, -\infty < s < t < \infty\}$ be a family of seminorms on $\mathcal{L}$ satisfying the following conditions:

(1) For $f_1, f_2 \in \mathcal{L}$, if $f_1(\tau) = f_2(\tau)$ for $s < \tau \leq t$, then $\|f_1 - f_2\|_{s,t} = 0$.
(2) Let L_τ denote shift to the left by τ. For all $f \in \mathcal{L}$, $\|L_\tau f\|_{s-\tau,t-\tau} = \|f\|_{s,t}$.
(3) Let $r < s < t$. Then for all $f \in \mathcal{L}$, $\|f\|_{s,t} \leq \|f\|_{r,t}$.
(4) Let $r < s < t$. Then for all $f \in \mathcal{L}$, $\|f\|_{r,t} \leq \|f\|_{r,s} + \|f\|_{s,t}$.
(5) There exists $0 < \alpha \leq \infty$ and $\mathcal{K} \geq 1$ such that if $0 < t - r \leq \alpha$ and $r < s < t$, then for

all $f \in \mathcal{L}$, $\|f\|_{r,s} \leq \mathcal{K}\|f\|_{r,t}$.

The pair $(\mathcal{L}, \mathcal{N})$ is called an FF *of seminorms* on $\mathcal{L}$. The normed linear space formed from equivalence classes of functions in $\mathcal{L}$ with norm $\|\cdot\|_{s,t}$ is denoted $H_{s,t}$. The elements of $H_{s,t}$ are the equivalence classes determined by: $f \sim g$, $f, g \in \mathcal{L}$ if and only if $\|f - g\|_{s,t} = 0$. They are denoted $u_{s,t}$, $y_{s,t}$, etc., i.e., the equivalence classes $u_{s,t}$, $y_{s,t}$ represent inputs and outputs, respectively. The set $\{H_{s,t}\}$, $-\infty < s < t < \infty$, is the *FF of normed linear spaces* given by $(\mathcal{L}, \mathcal{N})$.

A fairly wide class of examples of FFs is given by weighted L_p-spaces. For $1 \leq p < \infty$, let w be a fixed nonnegative, nonincreasing Lebesgue measurable real-valued function and let $\mathcal{L} = \mathcal{L}(\Re, \Re^N)$ be the set of N-vector-valued functions on $\Re$ that are p-integrable Lebesgue on finite intervals. Then, for $f \in \mathcal{L}$ the seminorms

$$\|f\|_{s,t} = \left(\int_s^t \|f(\tau)\|^p \, w(t-\tau) d\tau\right)^{1/p} \tag{2.1}$$

satisfy Conditions (1),...,(5) of Definition 1. For $p = \infty$ let $\mathcal{L}$ be the set of essentially bounded functions. When $\alpha = +\infty$ and $\mathcal{K} = 1$, $(\mathcal{L}, \mathcal{N})$ is a *standard* FF of seminorms. As time weighting was not essential to its purpose, standard FFs were used in [15].

We need to define another FF of seminorms. For $f \in \mathcal{L}$, put

$$\|f\|^{s,t} \triangleq \sup_{s<\tau\leq t} \|f\|_{s,\tau}. \tag{2.2}$$

With $\mathcal{M}$ indicating this new set of norms, $(\mathcal{L}, \mathcal{M})$ is indeed an FF of seminorms, [11].

An FF $(\mathcal{L}, \mathcal{N})$ and $\{H_{s,t}\}$, $-\infty < s < t < \infty$, can be augmented to include $\|\cdot\|_{-\infty,t}$ by taking the limit $s \to -\infty$, since by (3) of Definition 1 $\|f\|_{s,t}$ is monotone nondecreasing as $s \to -\infty$ with t fixed. Let $\mathcal{L}_0 = \{f \in \mathcal{L} | \lim_{s\to-\infty} \|f\|_{s,t} < \infty, t \in \Re\}$. For $f \in \mathcal{L}_0$, define

$$\|f\|_t = \|f\|_{-\infty,t} \triangleq \lim_{s\to-\infty} \|f\|_{s,t} . \tag{2.3}$$

With the meaning of $(\mathcal{L}, \mathcal{N})$ thus extended, $\|\cdot\|_{s,t}$ is defined for $-\infty \le s < t < \infty$. The *left-expanded FF of seminorms* is thereby defined and is denoted $(\mathcal{L}_0, \mathcal{N})$. It still satisfies all the Conditions (1), ..., (5).

Define $\mathcal{L}_{00}$ to be the subset of $\mathcal{L}_0$ such that for $f \in \mathcal{L}_{00}$ $\sup_t \|f\|_t < \infty$. Note that $\mathcal{L}_{00}$ is a translation-invariant linear space. For $f \in \mathcal{L}_{00}$, define

$$\|f\|_{s,\infty} \triangleq \sup_{t>s} \|f\|_{s,t} \;\; ; -\infty \le s . \tag{2.4}$$

It may be readily verified that if $(\mathcal{L}, \mathcal{N})$ is an FF for indices satisfying $-\infty < s < t < \infty$ then, with definitions given by (2.3) and (2.4), $(\mathcal{L}_{00}, \mathcal{N})$ is an FF for indices satisfying $-\infty \le s < t < \infty$ and satisfies conditions (1)–(3), and (5) of Definition 1 for indices $-\infty \le s < t \le \infty$. (For standard FFs, condition (4) holds for both cases.) Considering (2.2) and (2.4), writing $\|f\|^{s,\infty}$ is equivalent to $\|f\|_{s,\infty}$. $\{(\mathcal{L}_{00}, \mathcal{N}), \|\cdot\|_{s,t}, -\infty \le s < t \le \infty\}$ is called the *expanded family of seminorms* determined by $(\mathcal{L}, \mathcal{N})$. Note that $(\mathcal{L}_{00}, \mathcal{M})$ similarly defined is an FF for $-\infty \le s < t \le \infty$.

For $f \in \mathcal{L}_{00}$, we put

$$\|f\| \triangleq \sup_{t\in\Re} \|f\|_t = \|f\|_{-\infty,\infty} . \tag{2.5}$$

The normed linear space consisting of equivalence classes of functions in $\mathcal{L}_{00}$ with the norm (2.5) is called the *bounding space* H for the family $\{H_{s,t}\}$.

The *extended space* H^e for the family $\{H_{s,t}\}$ is the set of all equivalence classes of functions f in $\mathcal{L}_0$ ($f \sim g$ iff $\|f - g\| = 0$) for which $\|f\|_s < \infty$ for all s. It does not have a norm and, indeed, is given no topology. This definition agrees with the notion of extended space commonly used in the control literature.

It is possible that an FF $(\mathcal{L}, \mathcal{N})$ has a vacuous expansion in the sense that $\mathcal{L}_{00}$ is the empty set. An obvious example of this is given when $\mathcal{L}$ is the set of all constant real-valued functions on $\Re$ and $\mathcal{N}$ is the set of L_1-norms on finite intervals. To prevent this from happening and to further prevent the bounding space H from being too small (in a sense to be made explicit below), we can require that an FF be "full," as indicated in the following definition.

Definition 2. The FF $(\mathcal{L}, \mathcal{N})$ is *full* if each equivalence class $u_{s,t} \in H_{s,t}$, $-\infty < s < t < \infty$, has a representing function belonging to $\mathcal{L}_{00}$.[3]

When this definition is satisfied, then for all pairs (s,t), $-\infty < s < t < \infty$ there is a 1:1 correspondence between the normed linear space $H_{s,t}$ determined by $(\mathcal{L}, \mathcal{N})$ and the normed linear space $H'_{s,t}$ determined by $(\mathcal{L}_{00}, \mathcal{N})$, which preserves the normed linear space structure. The correspondence is given by $u_{s,t} \leftrightarrow u'_{s,t}$, $u_{s,t} \in H_{s,t}$, $u'_{s,t} \in H'_{s,t}$ if and only if $u_{s,t}$ and $u'_{s,t}$ have a common representing function $f \in \mathcal{L}_{00}$. Thus, if $(\mathcal{L}, \mathcal{N})$ is full, we need not distinguish between $H_{s,t}$ and $H'_{s,t}$. Henceforth, every FF mentioned is assumed to be full. The FFs formed with L_p spaces as described above arc full.

To emphasize the relations among the equivalence classes, suppose the function $f \in \mathcal{L}_{00}$ determines the equivalence classes $u \in H$, $u^t \in H^t$, $u_t \in H_t$, and $u_{s,t} \in H_{s,t}$. Since

$$\|f-g\|_{s,t} \le \|f-g\|_t \le \|f-g\|^t \le \|f-g\| \ ,$$

the equivalence class u considered as a set of functions is entirely contained in the equivalence class u^t considered as a set of functions, and similarly $u^t \subset u_t$ and $u_t \subset u_{s,t}$; also $u \subset u^{t,\infty}$. Thus, for example, given t, u determines u_t and $u^{t,\infty}$. Therefore, if f determines u and $-\infty \le s < t \le \infty$ it is meaningful, for example, to write $\|f\|_{s,t}$, $\|u\|_{s,t}$, $\|u^t\|_{s,t}$, $\|u^{s,\infty}\|_{s,t}$, $\|u_t\|_{s,t}$, $\|u_{s,t}\|_{s,t}$, and they are all equal. Let $-\infty \le r < s < t \le \infty$. Then, since $\|f\|_{s,t} \le \|f\|_{r,t}$ for $f \in \mathcal{L}_{00}$, the partitioning of $\mathcal{L}_{00}$ into equivalence classes by $\|\cdot\|_{r,t}$ results in a finer partition than that given by $\|\cdot\|_{s,t}$. That is, letting f determine $u \in H$, we have $u \subset u_{r,t} \subset u_{s,t}$.

It is sometimes necessary to consider an arbitrary past input concatenated with an arbitrary future input, i.e., to "splice" two inputs.

Definition 3 ([12]). For $-\infty \le r < s < t < \infty$, and $h, g \in \mathcal{L}$, the *splice* of h and g over $(r,t]$ at s is defined and equals f if

$$f(\tau) = \begin{cases} h(\tau), & r < \tau \le s \\ g(\tau), & s < \tau \le t \end{cases}$$

belongs to $\mathcal{L}$. It is denoted $f_{r,t} = h_{r,s} \leftrightarrow g_{s,t}$. For $-\infty \le r < s < \infty$, the splice of h and g over $(r,\infty]$ at s equals f if

$$f(\tau) = \begin{cases} h(\tau), & r < \tau \le s \\ g(\tau), & s < \tau \end{cases}$$

belongs to $\mathcal{L}$. It is denoted $f_{r,\infty} = h^{r,s} \leftrightarrow g_{s,\infty}$.

[3] This definition is due to Professor Root.

If $u_{r,s} \in H_{r,s}$, $v_{s,t} \in H_{s,t}$ are determined by functions h and g, respectively, and $h_{r,s} \overset{}{\longleftrightarrow} g_{s,t}$ exists, then for an FF, the splice of u and v (or $u_{r,s}$ and $v_{s,t}$) over $(r,t]$ at s is defined to be the element $w_{r,t} \in H_{r,t}$ determined by $h_{r,s} \longleftrightarrow g_{s,t}$; we write $w_{r,t} = u_{r,s} \longleftrightarrow v_{s,t}$. For $t = \infty$ we write $w_{r,\infty} = u^{r,s} \longleftrightarrow v_{s,\infty}$. These are not meaningful until it is proved that the splice is independent of the particular functions h and g representing the equivalence classes $u_{r,s}$ and $v_{s,t}$. However, this proof follows easily from Definition 1.

Any input space U is herein taken to be either the bounding space H of an FF $\{H_{s,t}\}$ that permits splicing or a translation-invariant subset of H. The extended space H^e can appear in an auxiliary role. We write $U_{s,t}$, $-\infty \le s < t \le \infty$, to denote the set ("space") of equivalence classes of functions belonging to $\mathcal{L}_{00}$ as determined by $\|\cdot\|_{s,t}$. If $U = H$, then $U_{s,t}$ is the normed linear space $H_{s,t}$; if U is a subset of H, $U_{s,t}$ is a space only in the sense that it is a subset of $H_{s,t}$. We call any $U_{s,t}$ a *truncated input space* and write U_t for $U_{-\infty,t}$.

The requirement that H permits splicing means that, if $U = H$, future inputs at any t can be arbitrary, with no regard to the past. Unfortunately, spaces of functions everywhere continuous on $\mathfrak{R}$ do not qualify, but this appears to be a minor drawback. It is sometimes desired that U be a translation-invariant bounded (or even totally bounded) subset of H; we always assume U contains the zero function. When U is a proper subset of H, a splice of two elements in U does not necessarily belong to U. However, we require that for all $u \in U$, both $u^t \longleftrightarrow 0_{t,\infty}$ and $0^t \longleftrightarrow u_{t,\infty}$ belong to U.

The output space Y is taken to be the bounding space, here denoted K, of an FF of normed linear spaces $\{K_{s,t}\}$, or occasionally the corresponding extended space K^e. In general, the families $\{H_{s,t}\}$ and $\{K_{s,t}\}$ need not be the same. The notations for output spaces are analogous to those for input spaces. The comments about equivalence classes are valid for the $y_{s,t} \in K_{s,t}$.

A mapping $F : U \to Y$ is called a *global input–output mapping* (or usually just an input–output mapping).

Definition 4. Let (Y, F, U) be an input–output system. F is a *causal mapping* and (Y, F, U) is a *causal system* if and only if for all t and for all $u, v \in U$ such that $\|u - v\|_t = 0$ it follows that $\|F(u) - F(v)\|_t = 0$.

If F satisfies this definition, it determines a mapping from U_t into Y_t, denoted $\tilde{F}_t$, that satisfies $\left\|\tilde{F}_t u_t - (Fu)_t\right\|_t = 0$. We call $\tilde{F}_t$ a *truncated input–output mapping* and define the *centered truncated input–output mapping* $F_t : U_0 \to Y_0$ by $F_t(u_0) \triangleq \mathrm{L}_t \tilde{F}_t \mathrm{R}_t(u_0)$, where $\mathrm{R}_t \triangleq \mathrm{L}_{-t}$ is the right-shift by t. We assume that all systems in this book are causal. If F satisfies this definition, it is causal in the usual sense. However, since causality is defined in terms of FFs, systems which "look like" causal systems may not be. Memory can affect causality.

Definition 5. Consider a left-expanded FF $\{H_{s,t}\}; -\infty \le s < t < \infty$. The family $\{H_{s,t}\}$ and the norms $\|\cdot\|_t$ are said to be *finite memory with memory length M* if there exists $0 < M < \infty$ such that $\|f\|_t = \|f\|_{t-M,t}$ for all $f \in L_0, t \in \mathfrak{R}$.

We see that if w in (2.1) has finite support, i.e., $w(t) = 0$ for $t \geq M$, the weighted L_p normed linear spaces are examples of finite memory spaces with memory length M. Finite memory may cause a system that is causal in the usual sense to not be causal according to Definition 4 because of the nature of the FF defined on the input space rather than the nature of the system itself. This does not happen for any FF where $\|u\|_t = 0$ implies $\|u\|_s = 0$ for all $s \leq t$—the usual case.

The next step in setting up mathematical structure is to specify norms to be used for input–output mappings $\tilde{F}_t$ and F_t. Although Lipschitz norms might seem at first to be the obvious choice, we much prefer, for reasons mentioned below, to use what we call N-power norms, introduced in [13] (see also [16]). The N-power norms, denoted $\|\cdot\|_{(N)}$, are defined as follows. Let Φ be a mapping from a normed linear space X into a normed linear space Y. For any nonnegative integer N, a norm for Φ is given by

$$\|\Phi\|_{(N)} \triangleq \sup_{x\in X} \frac{\|\Phi(x)\|}{1+\|x\|^N} \tag{2.6}$$

when the right side exists. (We omit the subscript (N) when possible). We say Φ is *bounded (in N-power norm)* if $\|\Phi\|_{(N)} < \infty$. If Φ is bounded, it carries bounded sets into bounded sets by the inequality

$$\|\Phi(x)\| \leq \|\Phi\|_{(N)} \cdot (1+\|x\|^N).$$

However, boundedness of Φ does not in general imply continuity, nor vice versa. Other properties of these norms are given in the appendix of [15] and in Appendix A of [16]. Although [15] uses the standard FFs, its appendix also applies to FFs in general. There is also a comparison of N-power norms and Lipschitz norms in these appendices. We have chosen to use N-power norms rather than Lipschitz norms because they are less restrictive and they are not as influenced by the fine structure of a mapping. This last property has relative importance when dealing with an approximate system representation, see [6], page 785.

Using (2.6) on the truncated system mapping, we have

$$\|F_t\|_{(N)} = \sup_{u_0} \frac{\|F_t(u_0)\|_0}{1+\|u_0\|_0^N} = \sup_{u_t} \frac{\|\tilde{F}_t(u_t)\|_t}{1+\|u_t\|_t^N} = \|\tilde{F}_t\|_{(N)}\,. \tag{2.7}$$

The norm for the global input–output mapping may be defined by

$$\|F\|^* = \sup_u \frac{\|F(u)\|}{1+\|u\|^N}. \tag{2.8}$$

However, in this book, input–output systems are themselves used as inputs to the generalized adjoint systems. Therefore, the norm for the global input–output mapping should be similar to the norm for an input. In this book, the global input–output mapping is defined by

$$\|F\| = \sup_t \|F_t\|_0 \triangleq \sup_t \|F_t\|_{-\infty,0}, \tag{2.9}$$

[16]. There is further elaboration about two global system norms (2.8) and (2.9) in the appendix.

The causal bounded input–output mappings are denoted by $\mathcal{D}_N(U, Y)$. The following lemma (a special case of Lemma A.1 in [16]) gives conditions so $\mathcal{D}_N(U, Y)$ is a Banach space.

Lemma 6 ([16]). *Let $Y_{-b,0}$ and Y be Banach spaces where $0 \leq b \leq \infty$. Then $\mathcal{D}_N(U_0, Y_{-b,0})$ is a Banach space with norm $\|\cdot\|_{-b,0}$ and $\mathcal{D}_N(U, Y)$ is a Banach space with norm $\|\cdot\|$, respectively.*

Definition 7 ([16]). An input–output system (Y, F, U), $F \in \mathcal{D}_N(U, Y)$ is *compatibly continuous* if the truncated maps F_t are equicontinuous.

Compatible continuity is stronger than continuity (it implies continuity). However, for a linear system with norm defined by (2.9), compatible continuity is equivalent to boundedness. Denote the set of causal bounded, compatibly continuous input–output systems by $\mathcal{C}_N(U, Y)$. It may be shown that if $\alpha = +\infty$, $0 \in U$, and U is closed with respect to splicing, compatible continuity is equivalent to continuity [16]. These results are derived in the appendix (Lemmas 39–41.)

Throughout this book, whenever there is reference to a system (Y, F, U), the following three hypotheses are in effect unless otherwise specifically noted:

(A) The input space U is either the bounding space H of an FF of normed linear spaces that permits splicing or a shift-invariant subset of such an A. If U is a proper subset of H, we require that it contain 0, but also that $u \in U$ implies both $u^t \leftrightarrow 0_{t,\infty}$ and $0^t \leftrightarrow u_{t,\infty}$ belong to U.
(B) The output space Y is the bounding space K of an FF of normed linear spaces.
(C) The global system operator F satisfies Definition 4 (causality) with respect to the given H and K.

The next hypothesis will often be needed but will not be in effect unless stated explicitly.

(D) The operators $F_t : U_0 \to Y_0$ are uniformly bounded in N-power norm for some fixed positive integer N by a constant C for all $t \in \Re$, and are an equicontinuous family of uniformly continuous mappings.

Hypothesis (D) implies that the global system operator is bounded with bound C and uniformly continuous (see Lemmas 4 and 5 of [15]).[4] Other properties requiring Hypothesis (D), having to do with the natural state, are given in [15, 16, 21].

---

[4]Or, $F \in \mathcal{C}_N(U, Y)$ and is uniformly continuous. The proofs given in [15] were intended for a standard FF; however, such proofs usually hold for FFs in general if Condition (5) of Definition 1 is not used.

$$[illegible] \, \|T\| = \sup [illegible] \quad (1.9)$$

[illegible] (1.9) [illegible] global [illegible] Lemma 1.2 [illegible] the appendix.

The causal bounded input–output mappings are denoted by [illegible]. The following lemma [illegible] of Lemma A.1 [illegible] conditions [illegible] is [illegible].

Lemma [illegible] [illegible] complete.

Definition 1.4 [illegible] system (X, F, G) [illegible] continuous.

[illegible] gain [illegible] complexity [illegible] systems with [illegible] of [illegible] compact [illegible] bounded [illegible] input–output systems [illegible] with respect to [illegible] These results are derived in the appendix [illegible] Lemma [illegible] there is [illegible] system (X, F, G) [illegible]

(A) The state space X is [illegible] space [illegible] subset [illegible]

(B) The [illegible] space Y is the [illegible]

[illegible] the [illegible] and [illegible]

The [illegible] will [illegible]

(C) The operator [illegible] on [illegible]

[illegible] continuity [illegible]

[illegible]

Chapter 3
Spaces of Time Functions Consisting of Input–Output Systems

The input space of time functions for a generalized adjoint system is discussed in this chapter. The time functions in this space are constructed from normalized time-truncated system trajectories. Let Z and Z_0 be Banach spaces. Then, using Lemma 6, $\mathcal{C}_N(Y, Z)$ and $\mathcal{C}_N(Y_0, Z_0)$ are Banach spaces. Let $\mathcal{B} = \{(Z, B, Y)\}$ be a Banach subspace of $\mathcal{C}_N(Y, Z)$. A truncated system $B_t : Y_0 \to Z_0$ is induced by each $t \in \Re$. (The systems in $\mathcal{B}$ are causal.) Therefore, the system trajectory $t \to B_t$ may be considered as a time function from $\Re$ to $\mathcal{C}_N(Y_0, Z_0)$. A linear space of time functions $\mathcal{L}(\Re, E_{Y^a})$, where $E_{Y^a} = \mathcal{C}_N(Y_0, Z_0)$, consisting of the time functions $t \to B_t$ for all $B \in \mathcal{B}$ may be defined by using the addition and scalar multiplication of input–output systems in $\mathcal{C}_N(Y, Z)$. Consider the time function $t \to B_t$ in the space $\mathcal{L}(\Re, E_{Y^a})$. The *left-translate* by T is $t \to B_{t+T}$. This defines an input–output system $B' = L_T B R_T$. Denote this translation by L_T^*. Then $L_T^* B = B' = L_T B R_T$. It is assumed that $\mathcal{B}$ is closed with respect to translations.

Concatenating one input with another is often a part of our calculations. The splice, defined in Chap. 2 (Definition 3), handles this. For example, often an input is truncated and spliced with the zero time function. We want to apply the splice concept to the time functions considered in this chapter. The splice (Definition 3) of $B^1, B^2 \in \mathcal{B}$ as time functions at $T \in \Re$ is the time function $\eta(t) \in \mathcal{L}(\Re, E_{Y^a})$ given by

$$\eta(t) = \begin{cases} B_t^1 & \text{for } t \le T \\ B_t^2 & \text{for } t > T \end{cases} . \tag{3.1}$$

The splice of B^1, B^2 will be permitted *only* if there exists $B \in \mathcal{B}$ such that $\eta(t) = B_t$ for all $t \in \Re$. A condition, which serves as a consistency condition, that ensures the splice of $B^1, B^2 \in \mathcal{C}_N(Y, Z)$ at time T is permitted is

$$\left\| B^2(y) - B^1(y) \right\|_T = 0 \tag{3.2}$$

D. Serakos, *Generalized Adjoint Systems*, SpringerBriefs in Optimization,
DOI 10.1007/978-3-319-16652-0_3

for all $y \in Y$. The splice of B^1, B^2 at time T, if it is permitted, e.g., (3.2) is satisfied, gives an input–output system $B \in \mathcal{C}_N(Y, Z)$ which is defined by

$$B(y) = ((B^1)^T \mapsto B^2_{T,\infty})(y) \triangleq (B^1(y))^T \mapsto (B^2(y))_{T,\infty} \,. \tag{3.3}$$

Note that even if B^1 and B^2 are time invariant, there is little chance that $(B^1)^T \mapsto B^2_{T,\infty}$ will be time invariant. Notice that in the case we are using standard fitted families for Z ($\alpha_Z + \infty$, $\mathcal{K}_Z = \infty$), (3.2) actually gives that $\left\| B^2(y) - B^1(y) \right\|_t = 0$ for all $t \le T$ and then (3.3) gives $(B^1)^T \mapsto B^2_{T,\infty} = B^2$. Usually, Eq. (3.2) is a sufficient condition.

Next, an FF of seminorms will be defined on $\mathcal{B}$. Since all $B \in \mathcal{B}$ are bounded in $\mathcal{C}_N(Y, Z)$, a seminorm $\| \cdot \|_{s,t}$ may be defined on $\mathcal{B}$ by

$$\| B \|_{s,t} \triangleq \sup_{y \in Y} \frac{\| B(y) \|_{s,t}}{1 + \| y \|_t^N}, \tag{3.4}$$

where $-\infty < s < t < +\infty$. In this book, $N \ge 1$; the case $N = 0$ is considered in [16]. The following proposition gives conditions for an FF of seminorms to be defined on $\mathcal{B}$.

Proposition 8. *If splicing is admissible in Y and $\alpha_Y = +\infty$, $\mathcal{K}_Y = 1$ (the variables α and $\mathcal{K}$ have been defined in Definition 1 and the subscripts will be used to identify them with a particular space), then $\mathcal{N}^1 \triangleq \{\| \cdot \|_{s,t}; -\infty < s < t < \infty\}$ is an FF of seminorms defined on the Banach space of causal input–output systems $\mathcal{B} = \{(Z, B, Y)\}$ viewed as time functions.*

Proof. Consider properties (1) through (5), which describe an FF of seminorms (see Definition 1).

(1) Let B^1, B^2 be such that $B^1_\tau = B^2_\tau$ for $s < \tau \le t$, then

$$\| B^1 - B^2 \|_{s,t} = \sup_{y \in Y} \frac{\left\| B^1(y) - B^2(y) \right\|_{s,t}}{1 + \| y \|_t^N} = 0.$$

(2) Let $T \in \Re$:

$$\| L_T^* B \|_{s-T,t-T} = \sup_{y \in Y} \frac{\| L_T B R_T(y) \|_{s-T,t-T}}{1 + \| y \|_{t-T}^N}$$

$$= \sup_{y \in Y} \frac{\| B R_T(y) \|_{s,t}}{1 + \| R_T y \|_t^N} = \| B \|_{s,t} \,.$$

(3) Let $r < s < t$:

$$\|B\|_{s,t} = \sup_{y \in Y} \frac{\|B(y)\|_{s,t}}{1 + \|y\|_t^N} \leq \sup_{y \in Y} \frac{\|B(y)\|_{r,t}}{1 + \|y\|_t^N} = \|B\|_{r,t}.$$

(4) Let $r < s < t$: Since $\alpha_Y = +\infty$, $\mathcal{K}_Y = 1$,

$$\begin{aligned}
\|B\|_{r,t} &= \sup_{y \in Y} \frac{\|B(y)\|_{r,t}}{1 + \|y\|_t^N} \\
&\leq \sup_{y \in Y} \frac{\|B(y)\|_{r,s}}{1 + \|y\|_t^N} + \sup_{y \in Y} \frac{\|B(y)\|_{s,t}}{1 + \|y\|_t^N} \\
&\leq \sup_{y \in Y} \frac{\|B(y)\|_{r,s}}{1 + \|y\|_s^N} \cdot \sup_{y \in Y} \frac{1 + \|y\|_s^N}{1 + \|y\|_t^N} + \|B\|_{s,t} \\
&\leq \|B\|_{r,s} + \|B\|_{s,t}.
\end{aligned}$$

(5) Let $r < s < t$ and $t - r < \alpha_Z$:

$$\begin{aligned}
\|B\|_{r,s} &= \sup_{y \in Y} \frac{\|B(y)\|_{r,s}}{1 + \|y\|_s^N} = \sup_{y_s \leftrightarrow 0_{s,t} \in Y_t} \frac{\|B(y_s \leftrightarrow 0_{s,t})\|_{r,s}}{1 + \|y_s \leftrightarrow 0_{s,t}\|_s^N} \\
&\leq \mathcal{K}_Z \sup_{y_s \leftrightarrow 0_{s,t} \in Y_t} \frac{\|B(y_s \leftrightarrow 0_{s,t})\|_{r,t}}{1 + \|y_s \leftrightarrow 0_{s,t}\|_t^N} \cdot \frac{1 + \|y_s \leftrightarrow 0_{s,t}\|_t^N}{1 + \|y_s \leftrightarrow 0_{s,t}\|_s^N} \\
&\leq \mathcal{K}_Z \sup_{y_t \in Y_t} \frac{\|B(y_t)\|_{r,t}}{1 + \|y_t\|_t^N} \cdot \sup_{y_s \leftrightarrow 0_{s,t} \in Y_t} \frac{1 + \|y_s \leftrightarrow 0_{s,t}\|_t^N}{1 + \|y_s \leftrightarrow 0_{s,t}\|_s^N} \\
&\leq \mathcal{K}_Z \|B\|_{r,t}.
\end{aligned}$$

Hence, $\alpha_{Y^a} = \alpha_Z$ and $\mathcal{K}_{Y^a} = \mathcal{K}_Z$. □

The seminorms $\|\cdot\|_t$ and $\|\cdot\|$ may be defined from $\|\cdot\|_{s,t}$ as indicated in Chap. 2. During calculations when Property (4) of Proposition 8 is not needed, the hypotheses $\alpha_Y = +\infty$, $\mathcal{K}_Y = 1$ are not applicable.

Chapter 4
A Generalized Adjoint System

In this chapter the generalized adjoint system will be defined and some of its properties investigated. We follow the arrangement of adjoints in the theory of linear operators. In the generalized adjoint systems functions, in some sense, are concatenated in reverse order. This gives the generalized adjoint of an input–output system the property of linearity, even though the (original) system is nonlinear. This is proven in the first proposition of the chapter, Proposition 11.

Let (Y, F, U) be a causal, possibly time-varying input–output system. Assume Y and Y_0 are Banach spaces so that, as a consequence of Lemma 6, $\mathcal{C}_N(U, Y)$ and $\mathcal{C}_N(U_0, Y_0)$ are Banach spaces with $N \geq 1$ (this will be needed in Chap. 5, where the generalized adjoint map and its invertibility are considered). (The case $N = 0$ is considered in [16].) Let $\mathcal{B} = \{(Z, B, Y\})$ be a translation-invariant closed linear subspace of causal, bounded input–output systems defined on the output space of F. Furthermore, let $\mathcal{B}$ be an input space of time functions as described in the previous chapter with $N = 1$, i.e., $B \in \mathcal{C}_1(Y, Z)$ for all $B \in \mathcal{B}$. By concatenating an input–output system $B \in \mathcal{B}$ to F, a causal input–output system $A = BF$ is defined. Note that A is an element of $\mathcal{C}_N(U, Z)$ where N is the same integer as is used for (Y, F, U) . Let $\mathcal{A}{=}\mathcal{C}_N(U, Z)$; then, $\mathcal{B}$ is the domain and $\mathcal{A}$ is the counter domain of the generalized adjoint system.

Since $\mathcal{A}$ and $\mathcal{B}$ are considered as spaces of time functions, we will need that they be FFs of normed linear spaces as described by Proposition 8. To verify Property (4) of the FFs, it was required in Proposition 8 that $\alpha_U = +\infty$, $\mathcal{K}_U = 1$ and $\alpha_Y = +\infty$, $\mathcal{K}_Y = 1$. (For bounded range systems, i.e., $(Y, F, U) \in \mathcal{C}_0(U, Y)$, this requirement is not necessary; see [16].) However, Property (4) is not used in the subsequent text; therefore, we relax these conditions.

The above conditions are in force throughout this book (unless otherwise specifically stated). In summary they are:

D. Serakos, *Generalized Adjoint Systems*, SpringerBriefs in Optimization,
DOI 10.1007/978-3-319-16652-0_4

1. $F \in \mathcal{C}_N(U, Y)$ and Y_0 and Y are Banach spaces, $N \geq 1$ ($N = 0$ in [16]).[1]
2. $B \in \mathcal{C}_1(Y, Z)$ for all $B \in \mathcal{B}$ and $\mathcal{B}$ is a closed, linear subspace of $\mathcal{C}_1(Y, Z)$ and Z_0 and Z are Banach spaces.
3. $\mathcal{A} = \mathcal{C}_N(U, Z)$.
4. $\mathcal{A}$ and $\mathcal{B}$ are translation invariant.

In the first proposition below concerning the generalized adjoint system, we will need this definition:

Definition 9. An input output system (Y, F, U) has memory length M if for all $-\infty \leq s < t < \infty$, for all $f, g \in U$, $\|f - g\|_{s-M,t} = 0$ implies $\|F(f) - F(g)\|_{s,t} = 0$. *An input–output system is memoryless if* $M = 0$.

A generalized adjoint system is defined as follows:

Definition 10. A *generalized adjoint system* (U^a, F^a, Y^a) of a causal input–output system (Y, F, U) has domain $Y^a = \mathcal{B}$ and counter domain $U^a = \mathcal{A}$ with F^a defined by $F^a : Y^a \to U^a$, $F^a(B) = A = BF$.

Proposition 11 gives some properties of generalized adjoint systems.

Proposition 11. *The generalized adjoint system F^a of a causal input–output system F is linear and memoryless (hence, causal). Also, F^a is time invariant if F is time invariant.*

Proof. LINEARITY: Given $B^1, B^2 \in \mathcal{B}$ and scalars α, β

$$\begin{aligned} F^a(\alpha B^1 + \beta B^2) &= (\alpha B^1 + \beta B^2)F \\ &= (\alpha B^1)F + (\beta B^2)F = \alpha(B^1 F) + \beta(B^2 F) = \alpha F^a(B^1) + \beta F^a(B^2). \end{aligned}$$

MEMORYLESS: Let $B^1, B^2 \in \mathcal{B}$ be such that $[\![B^1 - B^2]\!]_{s,t} = 0$ for $-\infty < s < t < \infty$ then,

$$\begin{aligned} [\![F^a(B^1) - F^a(B^2)]\!]_{s,t} &= \sup_u \frac{\|B^1 F(u) - B^2 F(u)\|_{s,t}}{1 + \|u\|_t^N} \\ &\leq \sup_u \frac{\|(B^1 - B^2)(Fu)\|_{s,t}}{1 + \|F(u)\|_t} \cdot \frac{1 + \|F(u)\|_t}{1 + \|u\|_t^N} \\ &\leq [\![B^1 - B^2]\!]_{s,t} \cdot (1 + [\![F]\!]_t) = 0. \end{aligned}$$

Hence, the system F^a is a memoryless system.

[1]Hypothesis (D) from Chap. 2 provides for $F \in \mathcal{C}_N(U, Z)$ in terms of conditions on the truncated systems. As stated in [15], it is better to have hypotheses in terms of the family $\{F_t\}$ rather than F because the mathematical description of a system is usually in terms of F_t.

TIME INVARIANCE: Assume F is time invariant: Consider

$$L_s^*(F^a(R_s^* B)) = L_s^*(F^a(R_s B L_s)) = L_s R_s B L_s F R_s = B L_s F R_s = BF = F^a(B),$$

where L_s^* is given by Property (2) in Proposition 8. Hence, the system F^a is a time invariant system. □

Since F^a is causal, a *truncated adjoint system* $\tilde{F}_t^a : Y_t^a \to U_t^a$ may be defined by

$$\tilde{F}_t^a(B_t) \triangleq (F^a B)_t = (BF)_t = B_t F_t,$$

where $B_t \in \mathcal{B}_t$ and $t \in \Re$. The *normalized truncated adjoint system* $F_a^t : Y_0^t \to U_0^a$ is defined by

$$F_t^a(B_0) \triangleq L_t^* \tilde{F}_t^a(R_t^* B_0) = L_t^* R_t^* B_0 F_t = B_0 F_t.$$

Since F_t^a is linear (shown similarly as for F^a in Proposition 11), the usual linear norm is used for F_t^a and $\| F^a \| = \sup_t \| F_t^a \|_0$, please refer to (2.9).

Proposition 12. *For $F \in \mathcal{C}_N(U, Y)$, F^a and F_t^a are bounded.*

Proof. F_t^a is uniformly continuous: Let $B_0^1, B_0^2 \in \mathcal{B}$ and recalling that by Condition 2 $\mathcal{B} \subseteq \mathcal{C}_1(Y, Z)$, then

$$\begin{aligned}
\| F_t^a(B_0^1) - F_t^a(B_0^2) \|_0 &= \| B_0^1 F_t - B_0^2 F_t \|_0 \\
&= \| (B_0^1 - B_0^2) F_t \|_0 = \sup_{u_0 \in U_0} \frac{\left\| (B_0^1 - B_0^2) F_t(u_0) \right\|_0}{1 + \|u_0\|_0^N} \\
&= \sup_{u_0 \in U_0} \frac{\left\| (B_0^1 - B_0^2) F_t(u_0) \right\|_0}{1 + \|F_t(u_0)\|_0} \cdot \frac{1 + \|F_t(u_0)\|_0}{1 + \|u_0\|_0^N} \\
&\le \| B_0^1 - B_0^2 \|_0 \cdot (1 + \|F_t\|_0) \le \| B_0^1 - B_0^2 \|_0 \cdot (1 + \|F\|) .
\end{aligned} \tag{4.1}$$

Equation (4.1) shows that F_t^a are equicontinuous; hence, F^a is compatibly continuous (Definition 7) and F^a is bounded (see Lemma 40). □

Chapter 5
A Generalized Adjoint Map

Consider the Banach space of causal, bounded, compatibly continuous input–output systems $\mathcal{F} = \{(Y, F, U)\} \subset \mathcal{C}_N(U, Y)$, all of which have the same input space U and the same output space Y. Fix an input space of causal input–output systems $\mathcal{B} = \{(Z, B, Y)\}$ and assume the conditions listed in Chap. 4 hold. Define an *adjoint map* ϕ with domain $\mathcal{F}$ by $\phi(F) = F^a$.[1] Let the range of ϕ be denoted by $\mathcal{F}^a \triangleq \phi(\mathcal{F})$. All adjoint systems $F^a \in \mathcal{F}^a$ will have common domain Y^a and common counter domain U^a. Note that ϕ is dependent on $\mathcal{B}$, but is well defined when $\mathcal{B}$ is fixed. Let the norm on ϕ be given by

$$\|\phi\| \triangleq \sup_{F \in \mathcal{F}} \frac{\|\phi(F)\|}{1 + \|F\|} .$$

First, we show that ϕ is bounded and translation preserving. Recall that since F^a is linear and $\|F^a\| = \sup_t \|F_t^a\|_0 = \sup_{t,B_0} \|F_t^a(B_0)\|_0 / \|B_0\|_0$.

Proposition 13. $\|\phi\| \leq 1$.

Proof.

$$\|\phi\| = \sup_F \frac{\|\phi(F)\|}{1 + \|F\|} = \sup_F \frac{\|F^a\|}{1 + \|F\|}$$

$$= \sup_F \frac{\sup_t \sup_{B_0} \frac{\|F_t^a(B_0)\|_0}{\|B_0\|_0}}{1 + \|F\|} = \sup_F \frac{\sup_t \sup_{B_0} \frac{\|B_0 F_t\|_0}{\|B_0\|_0}}{1 + \|F\|}$$

[1] In case we had been considering adjoints in finite dimensional linear operators the adjoint map would be the map taking the conjugate transpose.

D. Serakos, *Generalized Adjoint Systems*, SpringerBriefs in Optimization,
DOI 10.1007/978-3-319-16652-0_5

$$\leq \sup_F \frac{\sup_t \sup_{B_0} \dfrac{(1 + \|F_t\|_0) \cdot \|B_0\|_0}{\|B_0\|_0}}{1 + \|F\|} = 1 .$$

□

Proposition 14. *ϕ preserves translation. (Note that the left-translate of F^a by T is $L_T^* F^a R_T^*$.)*

Proof. For $B \in Y^a$

$$\begin{aligned}
\phi(L_T^* F)B &= \phi(L_T F R_T)B \\
&= (L_T F R_T)^a B = B L_T F R_T = L_T R_T B L_T F R_T \\
&= L_T^*(R_T B L_T F) = L_T^*(F^a(R_T B L_T)) \\
&= L_T^*(F^a(R_T^* B)) = (L_T^* \phi(F) R_T^*)B;
\end{aligned}$$

hence, ϕ preserves translation. □

A sufficient condition for ϕ to be one to one is: For all distinct $y_1, y_2 \in Y$, there exists $B \in \mathcal{B}$ such that $B(y_1) \neq B(y_2)$. (Other sufficient conditions are given later.) Also, it may be seen that if all $B \in Y^a$ are linear, then ϕ is linear.

The Lipschitz norm of an operator ϕ is

$$\|\phi\|_{\text{Lip}} = \sup_{F^1, F^2 \in \mathcal{F}} \frac{\|\phi(F^1) - \phi(F^2)\|}{\|F^1 - F^2\|} , \tag{5.1}$$

see [3] Page 50 or Page 39 of [2]. We say an operator with finite Lipschitz norm is Lipschitz continuous.

Lemma 15. *If in addition to the aforementioned hypotheses all $B \in \mathcal{B}$ are Lipschitz continuous, ϕ is continuous.*

Proof. We have:

$$\|\phi(F^1) - \phi(F^2)\| = \|F^{1a} - F^{2a}\|$$

$$\sup_t \sup_{B_0 \in Y_0^a} \frac{\|(F_t^{1a} - F_t^{2a})B_0\|_0}{\|B_0\|_0}$$

$$= \sup_t \sup_{B_0} \frac{\|B_0 F_t^1 - B_0 F_t^2\|_0}{\|B_0\|_0}$$

$$= \sup_{t, B_0} \frac{\sup_{u_0} \dfrac{\left\|(B_0 F_t^1)(u_0) - (B_0 F_t^2)(u_0)\right\|_0}{1 + \|u_0\|_0^N}}{\|B_0\|_0}$$

$$\leq \sup_{t,B_0} \frac{\|B_0\|_0 \cdot \sup_{u_0} \frac{\left\|F_t^1(u_0) - F_t^2(u_0)\right\|_0}{1 + \|u_0\|_0^N}}{\|B_0\|_0}$$

$$= \|F^1 - F^2\|,$$

which is sufficient for the continuity of ϕ (and gives $\|\phi\|_{\mathrm{Lip}} = 1$). □

To close out this chapter, we consider two propositions under which the quotient in (5.1) is bounded below and these will lead to the invertibility of ϕ and the continuity of ϕ^{-1}. Besides invertibility being a desirable property for ϕ to have, it is needed in Chap. 6. For the first proposition, we essentially show ϕ is invertible if there is a $B \in \mathcal{B}$ that is invertible.

Proposition 16. *In addition to the hypotheses of this chapter, assume there exists causal $B^L \in \mathcal{B}$ such that $\forall y^1, y^2 \in Y$, $\left\|B_0^L(y_0^1 - y_0^2)\right\|_0 \geq m \left\|y_0^1 - y_0^2\right\|_0$, where $0 < m < \infty$ (say B^L is Lipschitz bounded below). Then $\|\phi\|_{\mathrm{Lip}}$ in (5.1) is bounded below.*

Proof. We have

$$\begin{aligned}
\|\phi(F^1) - \phi(F^2)\| &= \|F^{1a} - F^{2a}\| = \sup_t \sup_{B_0 \in Y_0^a} \frac{\|(F_t^{1a} - F_t^{2a})B_0\|_0}{\|B_0\|_0} \\
&\geq \sup_t \frac{\|(F_t^{1a} - F_t^{2a})B_0^L\|_0}{\|B_0^L\|_0} = \sup_t \frac{\|B_0^L(F_t^1 - F_t^2)\|_0}{\|B_0^L\|_0} \\
&= \sup_t \frac{\sup_{u_0} \frac{\left\|B_0^L(F_t^1 - F_t^2)(u_0)\right\|_0}{1 + \|u_0\|_0^N}}{\|B_0^L\|_0} \\
&\geq \frac{\sup_t m \cdot \sup_{u_0} \frac{\left\|(F_t^1 - F_t^2)(u_0)\right\|_0}{1 + \|u_0\|_0^N}}{\|B^L\|} = \frac{m \cdot \|F^1 - F^2\|}{\|B^L\|}.
\end{aligned} \tag{5.2}$$

Hence, $\|\phi\|_{\mathrm{Lip}} \geq m/\|B^L\|$. □

The next lemma describes an input–output system $B : Y \to Z$, which has a specialized property. A system is presented that has norm one and, for a specified input, has output equal to the norm of that input. This system leads to our second proposition for the invertibility of ϕ. This system is causal, linear, and bounded. The construction uses the Hahn–Banach theorem. Specify a $y^* \in Y$ and $z^* \in Z$. In general z^* is an equivalence class of time functions. Pick a time function

ζ^* which is an element of z^*, that is, $\zeta^* \in [z^*]$. For each $t \in \Re$, there exists a linear functional $f^t : Y_t \to \Re$ such that for all $t \| f^t \| = 1$ and such that $f^t(y_t^*) = \|y_t^*\|_t$ (see the Hahn–Banach theorem, [4]). Define an input–output system (Z, B^*, Y) by

$$(B^*(y))(t) \triangleq f^t(y_t)\zeta^*(t) \,. \tag{5.3}$$

Note that the output of B^*, as described by (5.3), is a time function; however, the time function, in turn, specifies an equivalence class of time functions that, in general, is an element of Z.

Lemma 17. *The input–output system B^* defined by (5.3) is a linear system. If Y has the property that $\|y\|_t = 0$ implies $\|y\|_s = 0$ for all $y \in Y$, t and $s < t$, (e.g., $\alpha_Y = +\infty$ and $\mathcal{K}_Y = 1$), it is causal, and if the "sup" norm (L_∞) norm is used on Z, it is a bounded input–output system.*

Proof. LINEARITY: Let $y^1, y^2 \in Y, k_1, k_2 \in \Re$. Then,

$$\begin{aligned}(B^*(k_1y^1 + k_2y^2))(t) &= f^t((k_1y^1 + k_2y^2)_t)\zeta^*(t) \\ &= (k_1 f^t(y_t^1) + k_2 f^t(y_t^2))\zeta^*(t) \\ &= k_1 B^*(y_t^1)(t) + k_2 B^*(y_t^2)(t) \,.\end{aligned}$$

Therefore,

$$B^*(k_1y^1 + k_2y^2) = k_1 B^*(y^1) + k_2 B^*(y^2) \,.$$

CAUSALITY: Take $y^1, y^2 \in Y$ such that $\left\|y^1 - y^2\right\|_t = 0$. We then have that $\left\|y^1 - y^2\right\|_s = 0$ for all $s \le t$. Let $\zeta^1 = B^*\left(y^1\right)$ and $\zeta^2 = B^*\left(y^2\right)$. This implies $\zeta^1(s) = \zeta^2(s)$ for all $s \le t$, and by property (1) of the FFs (see Definition 1), $\left\|[\zeta^1] - [\zeta^2]\right\|_t = 0$, where the square brackets indicate the element of Z specified by the time functions; hence, B^* is causal.
BOUNDEDNESS: The linear operator norm is used for B^*.

$$\begin{aligned}\| B^* \| &= \sup_t \| B_t^* \| \\ &= \sup_t \sup_{y_0} \frac{\|B_t^*(y_0)\|_0}{\|y_0\|_0} \\ &= \sup_{y_0} \sup_t \frac{\|f^t(y_0)\zeta^*(t)\|}{\|y_0\|_0} \le \sup_{y_0} \frac{\sup_t |f^t(y_0)| \sup_t \|\zeta^*(t)\|}{\|y_0\|_0} \\ &\le \sup_{y_0} \frac{\sup_t \| f^t \| \, \|y_0\|_0 \, \|\zeta^*\|}{\|y_0\|_0} \le \sup_{y_0} \frac{\|y_0\|_0 \, \|\zeta^*\|}{\|y_0\|_0} = \|\zeta^*\| \,.\end{aligned}$$

Hence, B^* is bounded. □

For the second proposition, we show the quotient in (5.1) is bounded below if elements of the form (5.3) are available in $\mathcal{B}$.

Proposition 18. *Under the conditions that there exists causal bounded linear $B^* \in \mathcal{B}$ of the kind given by (5.3) (including that the "sup" norm is used in Z) then for all F^1, F^2, we have $\|\phi(F^1) - \phi(F^2)\| \geq \|F^1 - F^2\|$.*

Proof. Consider

$$\|\phi(F^1) - \phi(F^2)\| = \|F^{1a} - F^{2a}\| = \sup_t \sup_{B_0 \in Y_0^a} \frac{\|(F_t^{1a} - F_t^{2a})B_0\|_0}{\|B_0\|_0}$$

$$= \sup_t \sup_{B_0 \in Y_0^a} \frac{\|B_0(F_t^1 - F_t^2)\|_0}{\|B_0\|_0} \geq \sup_t \frac{\|B_t^*(F_t^1 - F_t^2)\|_0}{\|B_t^*\|_0}$$

$$= \sup_t \frac{\sup_{u_0} \dfrac{\left\|B_t^*(F_t^1 - F_t^2)(u_0)\right\|_0}{1 + \|u_0\|_0^N}}{\|B_t^*\|_0}$$

for any particular $B^* \in \mathcal{B}$. Let B^* be of the form given in Lemma (17), $B^*(y)(t) \triangleq f^t(y_t) \cdot \zeta^*(t)$ with $\|f^t\| = 1$ and $\|\zeta^*(t)\| = 1$ for all t. Substitution in the previous equation gives

$$= \sup_t \frac{\sup_{u_0} \dfrac{\left\|f^t((F_t^1 - F_t^2)(u_0)) \cdot \zeta^*(t)\right\|}{1 + \|u_0\|_0^N}}{\sup_{y_0} \dfrac{\|f^t(y_0) \cdot \zeta^*(t)\|}{\|y_0\|_0}} = \sup_t \frac{\sup_{u_0} \dfrac{\left|f^t((F_t^1 - F_t^2)(u_0))\right|}{1 + \|u_0\|_0^N}}{\sup_{y_0} \dfrac{|f^t(y_0)|}{\|y_0\|_0}}$$

$$\geq \sup_t \frac{\sup_{u_0} \dfrac{\left|f^t((F_t^1 - F_t^2)(u_0))\right|}{1 + \|u_0\|_0^N}}{\sup_{y_0} \dfrac{\|f^t\| \cdot \|y_0\|_0}{\|y_0\|_0}} = \sup_t \sup_{u_0} \frac{\left|f^t((F_t^1 - F_t^2)(u_0))\right|}{1 + \|u_0\|_0^N}.$$

Now, for all $\epsilon > 0$, $\exists\, \bar{t}$ and υ such that

$$\|F^1 - F^2\| \leq \|F_{\bar{t}}^1 - F_{\bar{t}}^2\|_0 + \frac{\epsilon}{2} \leq \frac{\left\|(F_{\bar{t}}^1 - F_{\bar{t}}^2)\upsilon_0\right\|_0}{1 + \|\upsilon_0\|_0^N} + \epsilon\,.$$

Select (using the Hahn–Banach theorem, [4]) f^t ($\|f^t\| = 1$) such that $f^t((F^1 - F^2)_t\upsilon_0) = \left\|(F^1 - F^2)_t\upsilon_0\right\|_0$, so for this f,

$$\frac{\left|f^t((F_t^1 - F_t^2)(\upsilon_0))\right|}{1 + \|\upsilon_0\|_0^N} = \frac{\left\|(F^1 - F^2)_t\upsilon_0\right\|_0}{1 + \|\upsilon_0\|_0^N}$$

for all t. Then,

$$\|\phi(F^1) - \phi(F^2)\| + \epsilon \geq \frac{\left\|(F_{\bar{t}}^1 - F_{\bar{t}}^2)\upsilon_0\right\|_0}{1 + \|\upsilon_0\|_0^N} + \epsilon \geq \|F^1 - F^2\|.$$

Since ϵ is arbitrary,

$$\|\phi(F^1) - \phi(F^2)\| \geq \|F^1 - F^2\|. \tag{5.4}$$

□

Propositions 16 and 18 give that ϕ is one to one. Let $F^1, F^2 \in \mathcal{F}$ with $F^1 \neq F^2$, which is equivalent to $\|F^1 - F^2\| > 0$. Equation (5.2) or (5.4) implies $\|\phi(F^1) - \phi(F^2)\| > 0$, which means $\phi(F^1) \neq \phi(F^2)$ so ϕ is one to one. Propositions 16 and 18 also give that ϕ^{-1} is uniformly continuous. Consider the boundedness of ϕ^{-1}. Let $F = F^1$ and $F^2 = 0$. Using these propositions, we get

$$\frac{\|F\|}{1 + \|\phi(F)\|} < \begin{cases} \|B^L\|/m & \text{from (5.2)} \\ 1 & \text{from (5.4)} \end{cases}$$

which are bounds for ϕ^{-1}.

Both of these propositions require that the domain space $\mathcal{B}$ contain specific or specialized elements. Roughly speaking, ϕ maps two different functions, F^1 and F^2, to two different functions, F^{1a} and F^{2a}, if the domain $\mathcal{B}$ is sufficiently large. In this regard, the domain $\mathcal{B}$ may be viewed as a space of "test inputs."

Chapter 6
On Invertibility Using the Generalized Adjoint System

Referring to Fig. 1.1, we see that the adjoint system F^a flows in the opposite direction to F. From this, we are led to consider whether F^a inverts F in some sense. This is the topic of this chapter. The results of this chapter are divided into two sections. In the first section, an algebraic analysis is presented. That is, we consider if F^a is one to one and onto when F is, and vice versa. In the second section, a functional analysis is presented. Referring once again to Fig. 1.1, we see that F and F^a map from different domains and ranges. To work out an inverse we want these domains and ranges to coincide. We define the auxiliary map G, in a space of maps $\mathcal{G}$ which as described, behaves like F and has input space U^a and output space Y^a, the output and input spaces of F^a. The causality, boundedness, and continuity of G is considered. The extent that F^a is the inverse of G is considered. This covers the invertibility of F by F^a. Also, we consider the boundedness, continuity, and invertibility of the map from $\mathcal{F}$ to $\mathcal{G}$.

Invertibility for F and F^a

First, we assume that both F and F^a are invertible and find a relationship between the inverses. From the definitions

$$F^a(B) = A = BF . \tag{6.1}$$

However, since F is invertible,

$$AF^{-1} = B ;$$

hence,

$$(F^{-1})^a A = B . \tag{6.2}$$

D. Serakos, *Generalized Adjoint Systems*, SpringerBriefs in Optimization,
DOI 10.1007/978-3-319-16652-0_6

Substituting (6.1) into (6.2), we obtain

$$(F^{-1})^a F^a(B) = B \, . \tag{6.3}$$

Substituting (6.2) into (6.1), we obtain

$$F^a(F^{-1})^a A = A \, . \tag{6.4}$$

The conclusion from (6.3) and (6.4) is that $(F^a)^{-1} = (F^{-1})^a$. This is similar to Lemma 7, page 479 in [4], for linear operators on Banach spaces.

Next we consider, in the algebraic sense, the invertibility of F^a based on the invertibility of F. Meaning, we consider whether F^a is one to one and onto if F is one to one and onto. Similar analysis for the invertibility of F based on F^a is also considered.

Lemma 19. *F onto implies F^a is one to one.*

Proof. Let $B_1, B_2 \in Y^a$ such that $B_1 \neq B_2$, that is, $B_1(y) \neq B_2(y)$ for some $y \in Y$. Pick $u \in U$ such that $F(u) = y$. This is possible since F is assumed to be onto. Then, with $A_1 = F^a(B_1)$, $A_2 = F^a(B_2)$,

$$A_1(u) = [F^a(B_1)](u) = B_1 F(u) = B_1(y)$$
$$A_2(u) = [F^a(B_2)](u) = B_2 F(u) = B_2(y) \, .$$

Hence, $A_1 \neq A_2$ and F^a is one to one. □

Lemma 20. *F one to one implies F^a is onto.*

Proof. Fix some $A \in U^a$. For this A, consider u and $z = A(u)$. For this u, F specifies a unique $y \in Y$ by $F(u) = y$ (since F is one to one). Hence, a particular B is defined by $B(y) = z$. Note that the domain of this B is not necessarily all of Y (it may be a subset of Y). In other words,

$$B(y) = A(F^{-1}(y)) \; \forall \; y \in \text{Range}(F) \, .$$

This B has the property that $F^a(B) = A$, so F^a is onto. □

If F is algebraically invertible, that is, F is one to one and onto, then from Lemmas 19 and 20, we have that F^a is also invertible. We consider the reverse. The following lemma requires that $\mathcal{A}$ contains elements which distinguish elements of U, i.e., for $u_1 \neq u_2$ there exists A such that $A(u_1) \neq A(u_2)$. The system described by (6.5) in the next section fulfills this.

Lemma 21. *Assume that $\mathcal{A}$ contains distinguishing A for all $u_1, u_2 \in U$, such as given by (6.5) below, then F^a onto implies F one to one.*

Proof. Consider $u_1, u_2 \in U$ such that $u_1 \neq u_2$ and let $A : U \to Z$ be such that $A(u_1) \neq A(u_2)$. Since F^a is onto, there exists $B \in Y^a$ such that $F^a(B) = A$. Let $y_1 = F(u_1)$ and $y_2 = F(u_2)$.

$$[F^a(B)](u_1) = BF(u_1) = A(u_1) = B(y_1)$$
$$[F^a(B)](u_2) = BF(u_2) = A(u_2) = B(y_2)\,.$$

Therefore, $B(y_1) \neq B(y_2)$. However, this is not possible for any B unless $y_1 \neq y_2$; hence, F is one to one. □

The following lemma requires that $\mathcal{B}$ contains elements of the following form: Given (Y, F, U), there is a $B_F \in \mathcal{B}$ such that $B_F(y) = 0$ for all $y \in \text{Range}(F)$ and $B_F(y) = 1$ for all $y \in \text{Range}(F)^C$.

Lemma 22. *Assume $\mathcal{B}$ contains elements of the form B_F mentioned above and that F^a is one to one. Then, F is onto.*

Proof. Suppose F is not onto. Define $B_1 = B_F$ and $B_2 = 2 \cdot B_F$. Since F is not onto, $B_1 \neq B_2$. Consider, $A_1 = F^a(B_1)$ and $A_2 = F^a(B_2)$. It is seen that $A_1 = A_2$, indicating F^a is not one to one; a contradiction. This is similar to the last three lines in the proof of Lemma 7 on page 479 in [4] for linear operators on Banach spaces. [If $\text{Range}(F)^C$ always contains an open set, allowances could be made showing this result with all $B \in Y^a$ being continuous.] □

Lemmas 21 and 22 give the reverse results to Lemmas 19 and 20. The condition on $\mathcal{A}$ in Lemma 21 and on $\mathcal{B}$ in Lemma 22 are a form of completeness on the range and domain of F^a, respectively. Note that the two completeness conditions are imposed on the adjoint space.

On F^a As the Inverse of F

In this section, the extent to which F^a is the inverse of F is considered. Since the domain and range of F are not the same as, respectively, the range and domain of F^a, the first task of this section is to construct a new system whose domain and range are the range and domain of F^a, and that behaves like F. We add the following condition to those indicated in Chap. 4:

5. The "ess sup"-norm (L_∞ norm) is used on Z.

A particular A is constructed that has specialized properties. Consider a specific $z \in Z$; z is (in general) an equivalence class of time functions and may be denoted by $[z]$ for emphasis. Let $\zeta \in [z]$ be a specified element in $[z]$. For each $u^\# \in U$, specify an $A^\# \in U^a$ by

$$A^\#(u)(t) = \left\| u_t - u_t^\# \right\|_t \cdot \zeta(t)\,, \tag{6.5}$$

where $t \in \mathfrak{R}$. This system may be defined using the truncated systems:

$$\tilde{A}_t^{\#}(u_t)(t) = \left\|u_t - u_t^{\#}\right\|_t \cdot \zeta(t)\,; \tag{6.6}$$

$$A_t^{\#}(u_0)(t) = \left\|u_0 - \left(L_t u^{\#}\right)_0\right\|_0 \cdot \zeta(t)\,. \tag{6.7}$$

Obviously $A^{\#}$ is not linear. The set of all such $A^{\#}$ is $\mathcal{A}^{\#} \subset \mathcal{A}$. We see that the map $\gamma : U \to \mathcal{A}^{\#}$ defined by $\gamma(u^{\#}) = A^{\#}$ is one to one and onto, provided $\zeta(t) \neq 0$. In fact, γ is onto by construction. To see that it is one to one, consider $u^{\#} \neq u^{\flat}$. In this case, there exists t such that $u_t^{\#} \neq u_t^{\flat}$. Hence, $\tilde{A}_t^{\#}(u_t^{\#})(t) = \left\|u_t^{\#} - u_t^{\#}\right\|_t \cdot \zeta(t) \neq \left\|u_t^{\#} - u_t^{\flat}\right\|_t \cdot \zeta(t) = \tilde{A}_t^{\flat}(u_t^{\#})(t)$. So $A^{\#} \neq A^{\flat}$, and γ is one to one.

The causality, boundedness, and continuity of $A^{\#}$ is considered in the next lemma.

Lemma 23. *$A^{\#}$ of the form given by (6.5) is causal, bounded, and uniformly compatibly continuous (see Definition 7).*

Proof. CONTINUITY AND CAUSALITY: Consider the Lipschitz norm of $A^{\#}$: Let $M \triangleq \sup_t \|\zeta(t)\|$,

$$\|A^{\#}\|_{\text{Lip}} \triangleq \sup_t \|\tilde{A}_t^{\#}\|_{\text{Lip},t} = \sup_t \sup_{u_1,u_2} \frac{\left\|\tilde{A}_t^{\#}(u_t^1) - \tilde{A}_t^{\#}(u_t^2)\right\|_t}{\left\|u_t^1 - u_t^2\right\|_t}$$

$$= \sup_{t,u_1,u_2} \frac{\left|\left\|u_t^1 - u_t^{\#}\right\|_t - \left\|u_t^2 - u_t^{\#}\right\|_t\right| \cdot \|\zeta(t)\|}{\left\|u_t^1 - u_t^2\right\|_t} \leq \sup_{t,u_1,u_2} \frac{\left\|u_t^1 - u_t^2\right\|_t}{\left\|u_t^1 - u_t^2\right\|_t} \cdot M = M\,.$$

The above, by Definitions 4 and 7, shows that $A^{\#}$ is causal and uniformly compatibly continuous.

BOUNDEDNESS: It is shown that $A^{\#}$ is bounded. Refer to (2.6) and (2.9):

$$\|A^{\#}\| \triangleq \sup_t \|\tilde{A}_t^{\#}\|_t = \sup_{t,u} \frac{\left\|\tilde{A}_t^{\#}(u_t)\right\|_t}{1 + \|u_t\|_t^N}$$

$$\leq \sup_{t,u} \frac{\left\|u_t - u_t^{\#}\right\|_t}{1 + \|u_t\|_t^N} \cdot \sup_t \|\zeta(t)\| \leq (1 + \sup_t \left\|u_t^{\#}\right\|_t) \cdot \sup_t \|\zeta(t)\|$$

$$= (1 + \sup_t \left\|u_t^{\#}\right\|_t) \cdot M = (1 + \left\|u^{\#}\right\|) \cdot M, \tag{6.8}$$

for any $N \geq 1$. □

In the following paragraphs, some properties of $\mathcal{A}^{\#}$ are considered.

Lemma 24. *If ζ is a constant, i.e., $\zeta(t) = \bar{\zeta}$ for all t, then $\mathcal{A}^{\#}$ is translation invariant.*

Proof. As discussed in the introductory paragraph of Chap. 3, the left-translate by τ of $A^\#$ is $L_\tau^* A^\# \triangleq L_\tau A^\# R_\tau$. Given $u^\#, u$, we have

$$\begin{aligned} L_\tau^* A^\#(u)(t) &= (L_\tau A^\# R_\tau)(u)(t) = L_\tau(A^\#(R_\tau u)(t)) \\ &= L_\tau(\|(R_\tau u)_t - u_t^\#\|_t \cdot \zeta(t)) = L_\tau(\|u_{t-\tau} - u_t^\#\|_t \cdot \zeta(t)) \\ &= L_\tau(\|u_{t-\tau} - u_t^\#\|_t) \cdot L_\tau \zeta(t) = \|u_t - u_{t+\tau}^\#\|_t \cdot \zeta(t+\tau) \,. \end{aligned} \tag{6.9}$$

Let $u^\flat = L_t u^\#$. Since U (the input space to F) is translation invariant $u^\flat \in U$. We have

$$L_\tau^* A^\# = A^\flat \in \mathcal{A}^\#,$$

where $A^\flat = \gamma(u^\flat)$. □

We consider the effect of a translation in U. As in the proof of Lemma 24, let $u^\flat = L_\tau u^\#$, then

$$\begin{aligned} A^\flat(u)(t) &= \|u_t - u_t^\flat\|_t \cdot \zeta(t) \\ &= \|u_t - (L_\tau u^\#)_t\|_t \cdot \zeta(t) = \|u_t - u^\#{}_{t+\tau}\|_t \cdot \zeta(t) \,. \end{aligned} \tag{6.10}$$

Using (6.9) we get

$$u^\# \xrightarrow{\gamma} A^\# \xrightarrow{L_\tau^*} (L_\tau^* A^\#) \,. \tag{6.11}$$

Using (6.10) we get

$$u^\# \xrightarrow{L_\tau} L_\tau u^\# = u^\flat \xrightarrow{\gamma} A^\flat \,. \tag{6.12}$$

Since (6.9) does not equal (6.10), in general, Eqs. (6.11) and (6.12) cannot be used to form a commutative diagram, i.e., γ does not preserve translation. However, if $\zeta = \overline{\zeta}$ a constant, then (6.11) and (6.12) do form a commutative diagram. Since the sup norm is used in Z, selecting ζ a constant is valid. We will designate $\mathcal{A}^\#$ as an input space and by Condition (4) listed above, it has to be translation invariant. Setting ζ as a constant fulfills Condition (4) of Chap. 4; however, since we will not be using this condition, we do not set ζ as a constant, which gives the subsequent presentation a little more generality.

Remark 25. $\mathcal{A}^\#$ is not closed with respect to addition, in general. In fact, for $u^\#$, $u^\flat \in U$ and $A^\#, A^\flat \in \mathcal{A}^\#$,

$$\begin{aligned} (A^\# + A^\flat)u(t) &= A^\# u(t) + A^\flat u(t) \\ &= [\|u_t - u_t^\#\|_t + \|u_t - u_t^\flat\|_t] \cdot \zeta(t) \,. \end{aligned}$$

The above is a causal input–output system from U to Z; however, it is not likely in $\mathcal{A}^{\#}$. Hence, $\mathcal{A}^{\#}$ is not shown to be closed with respect to addition.

Further properties of γ are considered. We observed in the above remark that γ is not, in general, linear. The next two lemmas examine the boundedness and continuity of γ and γ^{-1}.

Lemma 26. *Let* $\sup_t \|\zeta(t)\| = M < \infty$. *Then map* $\gamma : U \to \mathcal{A}^{\#}$ *is bounded and uniformly continuous.*

Proof. BOUNDEDNESS OF γ: Lemma 23 gives that

$$\|\gamma\| \triangleq \sup_{u^{\#}} \frac{\|\gamma(u^{\#})\|}{1+\|u^{\#}\|^N} = \sup_{u^{\#}} \frac{\|A^{\#}\|}{1+\|u^{\#}\|^N}$$
$$\leq \sup_{u^{\#}} \frac{(1+\|u^{\#}\|)\cdot \sup_t \|\zeta(t)\|}{1+\|u^{\#}\|^N} \leq 2\cdot M\ .$$

Hence, γ is bounded.

CONTINUITY OF γ:

$$\|\gamma(u^{\#})-\gamma(u^{\flat})\| = \|A^{\#}-A^{\flat}\| = \sup_t \|\tilde{A}^{\#}_t - \tilde{A}^{\flat}_t\|_t$$
$$= \sup_{t,u} \frac{\left\|\tilde{A}^{\#}_t(u_t)-\tilde{A}^{\flat}_t(u_t)\right\|_t}{1+\|u_t\|_t^N} = \sup_{t,u} \frac{\left\| \|u_t-u^{\#}_t\|_t \cdot \zeta(t) - \|u_t-u^{\flat}_t\|_t\cdot\zeta(t)\right\|}{1+\|u_t\|_t^N}$$
$$\leq \sup_t \left\|u^{\#}_t-u^{\flat}_t\right\|_t \cdot \sup_t \|\zeta(t)\| = \left\|u^{\#}-u^{\flat}\right\|\cdot M\ . \tag{6.13}$$

Hence, γ is uniformly continuous. □

Lemma 27. *Let* $\sup_t \|\zeta(t)\| = M < \infty$ *and* $\inf_t \|\zeta(t)\| = m > 0$. *Then the map* $\gamma^{-1} : \mathcal{A}^{\#} \to U$ *is bounded and continuous.*

Proof. CONTINUITY OF γ^{-1}:

$$\|\gamma(u^{\#})-\gamma(u^{\flat})\| = \|A^{\#}-A^{\flat}\| = \sup_t \|\tilde{A}^{\#}_t - \tilde{A}^{\flat}_t\|_t$$
$$\geq \sup_t \frac{\left\| \|u^{\flat}_t-u^{\#}_t\|_t \cdot \zeta(t) - \|u^{\flat}_t-u^{\flat}_t\|_t\cdot\zeta(t)\right\|}{1+\left\|u^{\flat}_t\right\|_t^N} \geq \sup_t \frac{\left\|u^{\flat}_t-u^{\#}_t\right\|_t}{1+\left\|u^{\flat}_t\right\|_t^N}\cdot \inf_t \|\zeta(t)\|$$
$$\geq \sup_t \frac{\left\|u^{\flat}_t-u^{\#}_t\right\|_t}{1+\left\|u^{\flat}_t\right\|_t^N}\cdot m \geq \left\|u^{\flat}-u^{\#}\right\| \cdot \frac{m}{1+\left\|u^{\flat}\right\|^N}\ . \tag{6.14}$$

Using this and (6.8), we have

$$m\cdot\left\|u^{\flat}\right\| \leq \|\gamma(0)-\gamma(u^{\flat})\| \leq \|\gamma(0)\| + \|\gamma(u^{\flat})\| \leq M + \|\gamma(u^{\flat})\|\ . \tag{6.15}$$

Hence, from (6.14),

$$\left\|u^\flat - u^\#\right\| \leq \frac{1 + \left\|u^\flat\right\|^N}{m} \cdot \|\gamma(u^\#) - \gamma(u^\flat)\|$$

$$\leq \frac{1 + \left(\frac{M + \|\gamma(u^\flat)\|}{m}\right)^N}{m} \cdot \|\gamma(u^\#) - \gamma(u^\flat)\| .$$

Hence, γ^{-1} is continuous. Note that (6.14) also gives that

$$\|\tilde{A}_t^\# - \tilde{A}_t^\flat\|_t \geq \left\|u_t^\flat - u_t^\#\right\|_t \cdot \frac{m}{1 + \left\|u^\flat\right\|^N}, \tag{6.16}$$

which will be needed below.

BOUNDEDNESS OF γ^{-1}: From (6.15),

$$\|\gamma(u^\#)\| + M \geq \left\|u^\#\right\| \cdot m .$$

Or

$$\overline{M} \cdot (\|\gamma(u^\#)\| + 1) \geq \left\|u^\#\right\| \cdot m,$$

where $\overline{M} = \max(M, 1)$. Therefore,

$$\frac{\left\|u^\#\right\|}{1 + \|\gamma(u^\#)\|} \leq \frac{\overline{M}}{m} \tag{6.17}$$

which gives $\|\gamma^{-1}\| \leq \overline{M}/m$. This is the norm $\|\gamma^{-1}\|_{(1)}$; however, as stated in the appendix of [15], if $M < N$, $\|\gamma^{-1}\|_{(N)} \leq 2 \cdot \|\gamma^{-1}\|_{(M)}$. □

In the next lemma, we consider whether γ is a monotonic function, i.e., if $\left\|u^{1\#}\right\| \leq \left\|u^{2\#}\right\|$, then $\left\|\gamma(u^{1\#})\right\| \leq \left\|\gamma(u^{2\#})\right\|$. Since orthogonality is needed in the proof, in this lemma we use a weighted L_2 norm on U.

The weighted $L_2(s,t)$ space (2.1), $-\infty < s < t < \infty$, with $p = 2$, is a Hilbert space with inner product

$$\langle u, v\rangle_{s,t} = \left(\int_s^t u(\tau)\overline{v(\tau)}w(t-\tau)d\tau\right)^{1/2} \tag{6.18}$$

when $w(t) > 0$ for all t. In Lemma 28, we assume the input space U is a Hilbert space with inner product (6.18).

Lemma 28. *If $\zeta(t) \neq 0$ for all t and the weighted L_2 norm derived from the inner product (6.18) is used on U with $w(t) > 0$ for all t, then γ is monotonic.*

Proof. Given $u^{\#}$, $u^{\flat}$, let $\left\|u^{\#}\right\|_0 < \left\|u^{\flat}\right\|_0$. Define a hyperplane P by $P \triangleq \{u \in U \mid \left\|u - u^{\#}\right\|_0 = \left\|u - u^{\flat}\right\|_0\}$. P divides U_0 into two regions. Region $U_0^{\#}$ contains $u_0^{\#}$ and region $U_0^{\flat}$ contains $u_0^{\flat}$. With this construction, $u_0 \in U_0^{\#}$ iff $\left\|u - u^{\#}\right\|_0 \leq \left\|u - u^{\flat}\right\|_0$ and $u_0 \in U_0^{\flat}$ iff $\left\|u - u^{\#}\right\|_0 \geq \left\|u - u^{\flat}\right\|_0$. Thus, $0_0 \in U_0^{\#}$. Next, consider the following quotients:

$$\frac{\left\|u - u^{\#}\right\|_0}{1 + \|u\|_0^N} \text{ and } \frac{\left\|u - u^{\flat}\right\|_0}{1 + \|u\|_0^N} . \tag{6.19}$$

For $u_0 \in U_0^{\#}$, the numerator of the right quotient is larger than the numerator of the left quotient. Next, consider $u_0 \in U_0^{\flat}$. Construct a perpendicular p from u_0 to P. Denote the point in $U_0^{\#}$ along p equidistant to P as u_0 by $\overline{u}_0$. Consider the quotients:

$$\frac{\left\|u - u^{\#}\right\|_0}{1 + \|u\|_0^N} \text{ and } \frac{\left\|\overline{u} - u^{\flat}\right\|_0}{1 + \|\overline{u}\|_0^N} . \tag{6.20}$$

We have that $\|\overline{u}\|_0 \leq \|u\|_0$ and that $\left\|u - u^{\#}\right\|_0 = \left\|\overline{u} - u^{\flat}\right\|_0$. Hence, in (6.20) the right quotient is larger than the left quotient.

Hence, from (6.19) and (6.20)

$$\sup_u \frac{\left\|u - u^{\#}\right\|_0}{1 + \|u\|_0^N} \leq \sup_u \frac{\left\|u - u^{\flat}\right\|_0}{1 + \|u\|_0^N} .$$

Therefore, if $A^{\#} = \gamma(u^{\#})$ and $A^{\flat} = \gamma(u^{\flat})$ such that $\left\|u^{\#}\right\| < \left\|u^{\flat}\right\|$, then

$$\begin{aligned} [\![A^{\#}]\!] = \sup_t [\![A_t^{\#}]\!]_t &= \sup_{t,u} \frac{\left\|u_t - u_t^{\#}\right\|_t \cdot \|\zeta(t)\|}{1 + \|u\|_t^N} \\ &\leq \sup_{t,u} \frac{\left\|u_t - u_t^{\flat}\right\|_t \cdot \|\zeta(t)\|}{1 + \|u\|_t^N} = \sup_t [\![A_t^{\flat}]\!]_t = [\![A^{\flat}]\!] \end{aligned}$$

and γ is monotonic. □

The input and output spaces U and Y of F are not the same as the input and output spaces Y^a and U^a of F^a. With this feature, F^a could not be the inverse of F. As stated in the first paragraph in this section of the chapter, an auxiliary system having the same input and output spaces as the output and input spaces of F^a and having similar behavior to F is needed. To define the auxiliary system, each $u^{\#} \in U$ is assigned to an $A^{\#} \in U^a$. Select a time function $\zeta \in [z] \in Z$, with $\zeta(t) \neq 0$, and define an $A^{\#} : U \to Z$ by $A^{\#} = \gamma\left(u^{\#}\right)$, Eq. (6.5). Also, for $y \in$ Range(F), define a $B^{\#} : Y \to Z$ by

$$B^{\#}(y)(t) \triangleq \left\|y_t - y_t^{\#}\right\|_t \cdot \zeta(t) . \tag{6.21}$$

Referring to (6.5), Eq. (6.21) defines $\gamma\left(y^{\#}\right) = B^{\#}$; γ works on Y the same as on U. The auxiliary map G of F has domain $\{A^{\#}\} \triangleq \mathcal{A}^{\#} \subset \mathcal{A}$ [see (6.5)] and is defined by

$$G(A^{\#})(y)(t) \triangleq \left\|y_t - \tilde{F}_t(u_t^{\#})\right\|_t \cdot \zeta(t)\,. \tag{6.22}$$

The map G is related to F in the following sense. If $y^{\#} = F(u^{\#})$ and $A^{\#} \in \mathcal{A}^{\#}$ and $B^{\#} \in \mathcal{B}^{\#}$ are derived from $u^{\#}$ and $y^{\#}$, respectively, as described above, then $G(A^{\#}) = B^{\#}$. Note that with domain $\mathcal{A}^{\#} \subset \mathcal{A}$, G has range $\mathcal{B}^{\#} \subset \mathcal{B}$. Also note that (6.22) defines a unique G for each F.

Next, we consider whether F^a inverts G. For this, assume that F is one to one. Let $y^{\#} = F(u^{\#})$ so that $B^{\#} = G(A^{\#})$. We have

$$\begin{aligned} F^a(B^{\#})u(t) &= (B^{\#}F)u(t) = B^{\#}(Fu)(t) \\ &= \left\|(F(u) - y^{\#})_t\right\|_t \cdot \zeta(t) = \left\|(F(u) - F\left(u^{\#}\right))_t\right\|_t \cdot \zeta(t)\,. \end{aligned} \tag{6.23}$$

Denote the set of input–output systems $F^a B$ of the form (6.23) by $\mathcal{A}_F^{\#}$. Define an equivalence relation "$\sim$" for systems in $\mathcal{A}^{\#}$ and $\mathcal{A}_F^{\#}$ based on null spaces, i.e., two systems are equivalent if their null spaces are exactly the same. Since F is assumed to be one to one, using this equivalence relation on the right-hand side of (6.23) we have

$$\sim \left\|\left(u - u^{\#}\right)_t\right\|_t \cdot \zeta(t) = (A^{\#})u(t)\,. \tag{6.24}$$

The systems given by (6.23) and (6.24) are the only two elements in their equivalence class. Hence, from (6.22) and (6.24),

$$F^a(G(A^{\#})) \sim A^{\#}\,. \tag{6.25}$$[1]

The subsequent results in this chapter handle the boundedness, continuity, and some other properties of the maps we have introduced. The next lemma considers elementary properties of the map G.

Lemma 29. *The operator G is memoryless. If $\sup_t \|\zeta(t)\| = M < \infty$ and $\inf_t \|\zeta(t)\| = m > 0$, G is bounded. If, in addition, F is compatibly continuous, then G is compatibly continuous.*

Proof. MEMORYLESS: Consider $A^{\#}, A^{\flat}$ such that $\|A^{\#} - A^{\flat}\|_{s,t} = 0$ for all $s, t \in \mathfrak{R}$. This gives that for all $u \in U$, $\left\|A^{\#}(u) - A^{\flat}(u)\right\|_{s,t} = 0$. From (6.5) (and since the "sup" norm is used on Z),

[1] An alternative to (6.5) and (6.22) are $A^{\#}(u)(t) = \delta(u_t - u_t^{\#}) \cdot \zeta(t)$ and $G(A^{\#})(y)(t) = \delta(y_t - \tilde{F}_t(u_t^{\#})) \cdot \zeta(t)$ where $\delta : U_t \to \mathfrak{R}$ is the delta function, that is $\delta(u_t) = 1$ if u_t is zero and $= 0$ else. This $A^{\#}$ is bounded, but not continuous; however, the equivalence relation in the invertibility calculation becomes an equality.

$$\left\| u_\tau - u_\tau^\# \right\|_\tau \cdot \zeta(\tau) - \left\| u_\tau - u_\tau^\flat \right\|_\tau \cdot \zeta(\tau) = 0 \tag{6.26}$$

for all $s < \tau \leq t$. Substituting $u = u^\#$ in (6.26) gives

$$\left\| u_\tau^\# - u_\tau^\flat \right\|_\tau \cdot \zeta(\tau) = 0 \text{ for all } s < \tau \leq t \,. \tag{6.27}$$

Now, for any $y \in Y$ we have

$$\begin{aligned} \left\| (G(A^\#) - G(A^\flat))(y)(\tau) \right\| &= \left\| \left\| y - F(u^\#) \right\|_\tau \cdot \zeta(\tau) - \left\| y - F(u^\flat) \right\|_\tau \cdot \zeta(\tau) \right\| \\ &\leq \left\| F(u^\flat) - F(u^\#) \right\|_\tau \cdot \| \zeta(\tau) \| = 0 \,, \end{aligned}$$

for all $s < \tau \leq t$, from (6.27) and the causality of F. The above implies

$$\| G(A^\#) - G(A^\flat) \|_{s,t} = 0 \,,$$

which indicates that G is memoryless (also causal). Therefore, one memoryless system inverts another memoryless system. One consideration is how a linear map, F^a, could be the inverse of a nonlinear map, G. However, note the input space of G is not a linear space. In fact, it is not closed with respect to addition (see Remark 25).

BOUNDEDNESS: Under the stated conditions, we show next that G is bounded.

$$\begin{aligned} \| G \| \stackrel{\Delta}{=} \sup_t \| G_t \|_t &= \sup_t \sup_{A^\# \in \mathcal{A}^\#} \frac{\| \tilde{G}_t(\tilde{A}_t^\#) \|_t}{1 + \| A_t^\# \|_t^N} \\ &= \sup_{A^\# \in \mathcal{A}^\#} \sup_{t,y} \frac{\left\| \tilde{G}_t(\tilde{A}_t^\#)(y_t) \right\|_t}{(1 + \| y_t \|_t)(1 + \| A_t^\# \|_t^N)} \\ &\leq \sup_{A^\# \in \mathcal{A}^\#} \sup_{t,y} \frac{\left\| y_t - \tilde{F}_t(u_t^\#) \right\|_t}{(1 + \| y_t \|_t)(1 + \| A_t^\# \|_t^N)} \cdot \sup_t \| \zeta(t) \| \\ &\leq \sup_{A^\# \in \mathcal{A}^\#} \sup_{t,y} \frac{\| y_t \|_t + \left\| \tilde{F}_t(u_t^\#) \right\|_t}{(1 + \| y_t \|_t)(1 + \| A_t^\# \|_t^N)} \cdot \sup_t \| \zeta(t) \| \\ &\leq \left(1 + \sup_{A^\# \in \mathcal{A}^\#} \sup_t \frac{\left\| \tilde{F}_t(u_t^\#) \right\|_t}{(1 + \| A_t^\# \|_t^N)} \right) \cdot M. \end{aligned} \tag{6.28}$$

A lower bound for $\| A_t^\# \|_t$ is needed. Consider

$$\begin{aligned} \| A_t^\# \|_t = \sup_u \frac{\left\| \tilde{A}_t^\#(u_t) \right\|_t}{1 + \| u_t \|_t^N} &= \sup_u \frac{\sup_{\tau \leq t} \left\| u_\tau - u_\tau^\# \right\|_\tau \cdot \| \zeta(\tau) \|}{1 + \| u_t \|_t^N} \\ &\geq \sup_u \frac{\left\| u_t - u_t^\# \right\|_t}{1 + \| u_t \|_t^N} \cdot \inf_{\tau \leq t} \| \zeta(\tau) \| \geq \left\| u_t^\# \right\|_t \cdot m \,. \end{aligned} \tag{6.29}$$

Substituting into (6.28) and using $\mu = \min(1, m)$ gives

$$\|G\| \leq \left(1 + \sup_t \sup_{u \in U} \frac{\left\|\tilde{F}_t(u_t)\right\|_t}{(1 + \|u_t\|_t^N)}\right) \cdot \frac{M}{\mu^N}$$
$$= \left(1 + \sup_t \|F_t\|\right) \cdot \frac{M}{\mu^N} = (1 + \|F\|) \cdot \frac{M}{\mu^N} . \tag{6.30}$$

CONTINUITY: Finally, we show that G is compatibly continuous under the stated conditions.

$$\|G(A^\#) - G(A^\flat)\|_t = \sup_y \frac{\left\|\tilde{G}_t(\tilde{A}_t^\#)(y_t) - \tilde{G}_t(\tilde{A}_t^\flat)(y_t)\right\|_t}{1 + \|y_t\|_t^N}$$
$$\leq \sup_y \frac{\left|\left\|y_t - \tilde{F}_t(u_t^\#)\right\|_t - \left\|y_t - \tilde{F}_t(u_t^\flat)\right\|_t\right|}{1 + \|y_t\|_t^N} \cdot \sup_t \|\zeta(t)\|$$
$$\leq \sup_y \frac{\left\|\tilde{F}_t(u_t^\#) - \tilde{F}_t(u_t^\flat)\right\|_t}{1 + \|y_t\|_t^N} \cdot M \leq \left\|\tilde{F}_t(u_t^\#) - \tilde{F}_t(u_t^\flat)\right\|_t \cdot M .$$

Since F is compatibly continuous, given $u^\flat$ for all $\epsilon > 0$, there exists $\delta > 0$ such that if $\left\|u_t^\# - u_t^\flat\right\|_t < \delta$ then $\left\|\tilde{F}_t(u_t^\#) - \tilde{F}_t(u_t^\flat)\right\|_t < \epsilon$, for all t. Hence, using (6.16), G is compatibly continuous. □

We consider the boundedness and continuity of the map $\rho : \mathcal{F} \to \mathcal{G}$ ($\rho(F) = G$).

Lemma 30. *Let* $\sup_t \|\zeta(t)\| = M < \infty$ *and* $\inf_t \|\zeta(t)\| = m > 0$. *Then map* $\rho : \mathcal{F} \to \mathcal{G}$ *is bounded and continuous.*

Proof. BOUNDEDNESS: Consider

$$\|\rho\| \triangleq \sup_F \frac{\|G\|}{1 + \|F\|} .$$

However, this is just given by (6.30), that is,

$$\|\rho\| \leq \frac{M}{(\mu)^N} , \tag{6.31}$$

where $\mu = \min(1, m)$.

CONTINUITY: Let $\rho\left(F^1\right) = G^1$ and $\rho\left(F^2\right) = G^2$. Consider

$$\|G^1 - G^2\| = \sup_t \|G_t^1 - G_t^2\|_t = \sup_{t,A} \frac{\|\tilde{G}_t^1(\tilde{A}_t^\#) - \tilde{G}_t^2(\tilde{A}_t^\#)\|_t}{1 + \|\tilde{A}_t^\#\|_t^N}$$

$$= \sup_{t,A,y} \frac{\left\| \tilde{G}_t^1(\tilde{A}_t^\#)(y_t) - \tilde{G}_t^2(\tilde{A}_t^\#)(y_t) \right\|_t}{(1 + \|y_t\|_t) \cdot (1 + \|\tilde{A}_t^\#\|_t^N)}$$

or

$$= \sup_{t,A,y} \frac{\left| \left\| y_t - \tilde{F}_t^1(u_t^\#) \right\|_t - \left\| y_t - \tilde{F}_t^2(u_t^\#) \right\|_t \right|}{(1 + \|y_t\|_t) \cdot (1 + \|\tilde{A}_t^\#\|_t^N)} \cdot \|\zeta(t)\|$$

$$\leq \sup_{t,A} \frac{\left\| \tilde{F}_t^1(u_t^\#) - \tilde{F}_t^2(u_t^\#) \right\|_t}{(1 + \|\tilde{A}_t^\#\|_t^N)} \cdot M$$

$$\leq \sup_{t,u} \frac{\left\| \tilde{F}_t^1(u_t) - \tilde{F}_t^2(u_t) \right\|_t}{(1 + \|u_t\|_t^N)} \cdot \frac{M}{\mu^N} \leq \|F^1 - F^2\| \cdot \frac{M}{\mu^N},$$

using (6.29). □

We consider the boundedness and continuity of the map $\rho^{-1} : \mathcal{G} \to \mathcal{F}$ ($\rho^{-1}(G) = F$).

Lemma 31. *Let* $\sup_t \|\zeta(t)\| = M < \infty$ *and* $\inf_t \|\zeta(t)\| = m > 0$. *Then map* $\rho^{-1} : \mathcal{G} \to \mathcal{F}$ *is bounded. With the coarser* $C_{(2\cdot N)}$ *topology on* $\mathcal{F}$, ρ^{-1} *is continuous.*

Proof. BOUNDEDNESS: Consider

$$\|F\| = \sup_t \|F_t\|_t = \sup_{t,u} \frac{\left\| \tilde{F}_t(u_t) \right\|_t}{1 + \|u_t\|_t^N} = \sup_{t,u} \frac{\left\| \tilde{F}_t(u_t) \right\|_t}{1 + \|A_t\|_t^N} \cdot \frac{1 + \|A_t\|_t^N}{1 + \|u_t\|_t^N},$$

where $A = \gamma(u)$,

$$\leq \sup_{t,u,y} \cdot \frac{\left\| y_t - \tilde{F}_t(u_t) \right\|_t \cdot \|\zeta(t)\|}{(1 + \|y_t\|_t) \cdot (1 + \|A_t\|_t^N) \cdot m} \cdot \sup_{t,u} \frac{1 + \|A_t\|_t^N}{1 + \|u_t\|_t^N}$$

$$\leq \sup_{t,u,y} \cdot \frac{\left\| \tilde{G}_t(\tilde{A}_t)(y_t) \right\|_t}{(1 + \|y_t\|_t) \cdot (1 + \|A_t\|_t^N) \cdot m} \cdot \sup_{t,u} \frac{1 + \|A_t\|_t^N}{1 + \|u_t\|_t^N}$$

$$= \sup_{t,A} \cdot \frac{\|\tilde{G}_t(\tilde{A}_t)\|_t}{(1 + \|A_t\|_t^N) \cdot m} \cdot \sup_{t,u} \frac{1 + \|A_t\|_t^N}{1 + \|u_t\|_t^N} = \frac{\|G\|}{m} \cdot \sup_{t,u} \frac{1 + \|A_t\|_t^N}{1 + \|u_t\|_t^N}, \tag{6.32}$$

where in (6.32) we recall from Lemma 23 that $B^\# \in C_1$. Continuing, we see that from (6.8)

$$\frac{1 + \|A_t\|_t^N}{1 + \|u_t\|_t^N} \leq \frac{1 + ((1 + \|u_t\|_t) \cdot M)^N}{1 + \|u_t\|_t^N} = c < \infty. \tag{6.33}$$

Using this in (6.32) we obtain

$$\|\rho^{-1}(G)\| = \|F\| \le \frac{c}{m} \cdot \|G\| \tag{6.34}$$

and this gives that ρ^{-1} is bounded.

CONTINUITY: Let $\rho^{-1}\left(G^1\right) = F^1$ and $\rho^{-1}\left(G^2\right) = F^2$. Consider

$$\begin{aligned}\|G_t^1 - G_t^2\|_t &= \sup_A \frac{\|\tilde{G}_t^1(\tilde{A}_t) - \tilde{G}_t^2(\tilde{A}_t)\|_t}{1 + \|A_t\|_t^N} \\ &= \sup_{A,y} \frac{\left\|\left(\left\|y_t - \tilde{F}_t^1(u_t)\right\|_t - \left\|y_t - \tilde{F}_t^2(u_t)\right\|_t\right) \cdot \zeta(t)\right\|_t}{(1 + \|y_t\|_t) \cdot (1 + \|A_t\|_t^N)}.\end{aligned}$$

Letting $y_t = \tilde{F}_t^2(u_t)$ in the above equation gives for all A, u

$$\ge \frac{\left\|\tilde{F}_t^2(u_t) - \tilde{F}_t^1(u_t)\right\|_t \cdot \|\zeta(t)\|}{(1 + \left\|\tilde{F}_t^2(u_t)\right\|_t) \cdot (1 + \|A_t\|_t^N)}. \tag{6.35}$$

There are two cases to consider in (6.35).

Case 1: $\left\|\tilde{F}_t^2(u_t)\right\| \le 1$. This gives that in (6.35)

$$\begin{aligned}&\ge \frac{\left\|\tilde{F}_t^2(u_t) - \tilde{F}_t^1(u_t)\right\|_t \cdot m}{(1 + \|u_t\|_t^N)} \cdot \frac{(1 + \|u_t\|_t^N)}{2 \cdot (1 + \|A_t\|_t^N)} \ge \frac{\left\|\tilde{F}_t^2(u_t) - \tilde{F}_t^1(u_t)\right\|_t}{(1 + \|u_t\|_t^N)} \cdot \frac{m}{2 \cdot c} \\ &\ge \frac{\left\|\tilde{F}_t^2(u_t) - \tilde{F}_t^1(u_t)\right\|_t}{(1 + \|u_t\|_t^{2 \cdot N})} \cdot \frac{m}{2 \cdot c} \cdot \frac{1}{2},\end{aligned} \tag{6.36}$$

where c is the constant in (6.33).

Case 2: $\left\|\tilde{F}_t^2(u_t)\right\| > 1$. This gives that in (6.35)

$$\begin{aligned}&\ge \frac{\left\|\tilde{F}_t^2(u_t) - \tilde{F}_t^1(u_t)\right\|_t \cdot m}{2 \cdot \left\|\tilde{F}_t^2(u_t)\right\|_t \cdot (1 + \|A_t\|_t^N)} \\ &\ge \frac{\left\|\tilde{F}_t^2(u_t) - \tilde{F}_t^1(u_t)\right\|_t}{(1 + \|u_t\|_t^N)^2} \cdot \frac{m \cdot (1 + \|u_t\|_t^N)}{\|F_t^2\|_t \cdot 2 \cdot (1 + \|A_t\|_t^N)} \\ &\ge \frac{\left\|\tilde{F}_t^2(u_t) - \tilde{F}_t^1(u_t)\right\|_t}{(1 + \|u_t\|_t^{2 \cdot N})} \cdot C \cdot \frac{1}{\|F_t^2\|_t}\end{aligned} \tag{6.37}$$

[using (6.33)] where C is a constant greater than zero. Combining (6.36) and (6.37), we get that

$$\|G_t^2 - G_t^1\|_{(N)t} \geq \|F_t^2 - F_t^1\|_{(2\cdot N)t} \cdot \min\left(\frac{1}{2\cdot c\cdot 2}, \frac{C}{\|F_t^2\|_t}\right).$$

Hence,

$$K \cdot \max(1, \|F_t^2\|_t) \cdot \|G_t^2 - G_t^1\|_{(N)t} \geq \|F_t^2 - F_t^1\|_{(2\cdot N)t},$$

for some $0 < K < \infty$. Using (6.34) we obtain

$$K' \cdot \max(1, \|G_t^2\|_t) \cdot \|G_t^2 - G_t^1\|_{(N)t} \geq \|F_t^2 - F_t^1\|_{(2\cdot N)t}. \tag{6.38}$$

For some $0 < K' < \infty$. Equation (6.38) gives that ρ^{-1} is continuous, with the $\mathcal{C}_N$ topology in $\mathcal{G}$ and the coarser $\mathcal{C}_{(2\cdot N)}$ topology in $\mathcal{F}$. From Lemma 35 in the appendix, if $F \in \mathcal{C}_N$ then $F \in \mathcal{C}_{(2\cdot N)}$. □

Denote a functional form of the equivalence operator given in (6.24) by D. That is,

$$D(\|F(u) - F(u^{\#})\| \cdot \zeta(t)) \triangleq \|u - u^{\#}\| \cdot \zeta(t).$$

We conclude this chapter by considering the boundedness and continuity of this operator and its inverse.

Lemma 32. *Let $F(0) = 0$ and F^{-1} be bounded and continuous. Let $B^{\#} = G(A^{\#})$ and $A^{\#} \sim F^a(B^{\#})$ as in (6.24). Then $D(F^a(B^{\#})) = A^{\#}$ so defined is bounded and continuous.*

Proof. BOUNDEDNESS: Consider

$$\begin{aligned}
\|D\| &= \sup_{B^{\#}} \frac{\|A^{\#}\|}{1 + \|F^a(B^{\#})\|^N} = \sup_{B^{\#}} \frac{\sup_{t,u} \dfrac{\left\|u_t - u_t^{\#}\right\|_t \cdot \|\zeta(t)\|}{1 + \|u_t\|_t^N}}{1 + \|F^a(B^{\#})\|^N} \\
&= \sup_{B^{\#}} \frac{\sup_{t,u} \dfrac{\left\|u_t - u_t^{\#}\right\|_t \cdot \|\zeta(t)\|}{1 + \|u_t\|_t^N}}{1 + \sup_t \|\tilde{F}_t^a(\tilde{B}_t^{\#})\|_t^N} \\
&\leq \sup_{t,B^{\#}} \frac{\sup_{u} \dfrac{\left\|u_t - u_t^{\#}\right\|_t \cdot \|\zeta(t)\|}{1 + \|u_t\|_t^N}}{1 + \|\tilde{F}_t^a(\tilde{B}_t^{\#})\|_t^N} \leq M + \sup_{t,u^{\#}} \frac{\left\|u_t^{\#}\right\|_t \cdot M}{1 + (m \cdot \left\|\tilde{F}_t(u_t^{\#})\right\|_t)^N} < \infty,
\end{aligned} \tag{6.39}$$

since F^{-1} is bounded and, referring to (6.23), we have

$$\|\tilde{F}_t^a(\tilde{B}_t^\#)\|_t \geq \sup_{u_t} \frac{\left\| \tilde{F}_t(u_t) - \tilde{F}_t(u_t^\#) \right\|_t}{1 + \|u_t\|_t^N} \cdot m \geq \left\| \tilde{F}_t(u_t^\#) \right\|_t \cdot m$$

with $F(0) = 0$.

CONTINUITY: Let $D(F^a(B^\#)) = A^\#$ and $D(F^a(B^\flat)) = A^\flat$. Recalling (6.13), we have

$$\|A^\# - A^\flat\| \leq \left\| u^\# - u^\flat \right\| \cdot M . \tag{6.40}$$

Now,

$$\|\tilde{F}_t^a(\tilde{B}_t^\#) - \tilde{F}_t^a(\tilde{B}_t^\flat)\|_t \geq \frac{\left\| \tilde{F}_t(u_t^\#) - \tilde{F}_t(u_t^\flat) \right\|_t}{1 + \left\| u_t^\# \right\|_t^N} \cdot m ,$$

or

$$\begin{aligned} \left\| \tilde{F}_t(u_t^\#) - \tilde{F}_t(u_t^\flat) \right\|_t &\leq \left(\frac{1 + \left\| u_t^\# \right\|_t^N}{m} \right) \cdot \|\tilde{F}_t^a(\tilde{B}_t^\#) - \tilde{F}_t^a(\tilde{B}_t^\flat)\|_t \\ &\leq \left(\frac{1 + (\|A_t^\#\|_t / m)^N}{m} \right) \cdot \|\tilde{F}_t^a(\tilde{B}_t^\#) - \tilde{F}_t^a(\tilde{B}_t^\flat)\|_t \end{aligned}$$

from (6.29). Taking the supremum over t on both sides gives

$$\left\| F(u^\#) - F(u^\flat) \right\| \leq \left(\frac{1 + (\|A^\#\| / m)^N}{m} \right) \cdot \|F^a(B^\#) - F^a(B^\flat)\| . \tag{6.41}$$

We see from (6.39) that for some C

$$\frac{\|A^\#\|}{1 + \|F^a(B^\#)\|^N} < C < \infty .$$

Therefore,

$$\|A^\#\| < C \cdot (1 + \|F^a(B^\#)\|^N) .$$

Substituting into (6.41) gives

$$\begin{aligned} &\left\| F(u^\#) - F(u^\flat) \right\| \\ &\quad \leq \left(\frac{1 + (C \cdot (1 + \|F^a(B^\#)\|^N) / m)^N}{m} \right) \cdot \|F^a(B^\#) - F^a(B^\flat)\| . \end{aligned} \tag{6.42}$$

Using (6.40), (6.42) and the continuity of F^{-1}, we conclude that D is continuous. □

Lemma 33. *Let $B^\# = G(A^\#)$ and $A^\# \equiv F^a(B^\#)$ as in (6.24). Then $D^{-1}(A^\#) = F^a(B^\#)$ so defined is bounded and continuous.*

Proof. BOUNDEDNESS: Consider

$$\|D^{-1}\| = \sup_{A^\#} \frac{\|F^a(B^\#)\|}{1+\|A^\#\|^N} \le \sup_{t,A^\#} \frac{\|\tilde{F}^a_t(\tilde{B}^\#_t)\|_t}{1+\|A^\#_t\|_t^N}$$

$$\le \sup_{t,A^\#} \frac{\sup_u \dfrac{\sup_{\tau\le t} \left\|\tilde{F}_\tau(u_\tau) - \tilde{F}_\tau(u^\#_\tau)\right\|_\tau \cdot \|\zeta(\tau)\|}{1+\|u_t\|_t^N}}{1+(m\cdot\left\|u^\#_t\right\|_t)^N},$$

where we use (6.23) and (6.29),

$$\le M\cdot\left(\|F\| + \sup_{t,u^\#} \frac{\left\|\tilde{F}_t(u^\#_t)\right\|_t}{1+(m\cdot\left\|u^\#_t\right\|_t)^N}\right) < \infty\,.$$

CONTINUITY: Let $D^{-1}(A^\#) = F^a(B^\#)$ and $D^{-1}(A^\flat) = F^a(B^\flat)$. Consider

$$\begin{aligned}\|F^a(B^\#) - F^a(B^\flat)\| &= \sup_t \|\tilde{F}^a_t(\tilde{B}^\#_t) - \tilde{F}^a_t(\tilde{B}^\flat_t)\|_t \\ &\le \sup_{t,u} \frac{\left|\left\|\tilde{F}_t(u_t) - \tilde{F}_t(u^\#_t)\right\| - \left\|\tilde{F}_t(u_t) - \tilde{F}_t(u^\flat_t)\right\|\right|\cdot M}{1+\|u_t\|_t^N} \\ &\le \sup_{t,u} \frac{\left\|\tilde{F}_t(u^\#_t) - \tilde{F}_t(u^\flat_t)\right\|_t}{1+\|u_t\|_t^N}\cdot M \le \left\|F(u^\#) - F(u^\flat)\right\|\cdot M\,.\end{aligned} \tag{6.43}$$

However, using Lemma 27

$$\left\|u^\# - u^\flat\right\| \le \|\gamma(u^\#) - \gamma(u^\flat)\| \cdot \frac{1+\left(\dfrac{M+\|A^\#\|}{m}\right)^N}{m};$$

therefore, using this, the continuity of F and (6.43), D^{-1} is continuous. □

Summary

In this chapter we analyzed the generalized adjoint system. The main theme is invertibility. Initially, we found that F^a is one to one when F is onto and F^a is onto when F is one to one. With some conditions, F is one to one when F^a is onto and F is onto when F^a is one to one.

The analysis continues from the functional viewpoint. The lower portion of Fig. 6.1 depicts the situation. Four equations give the essentials: Eqs. (6.5), (6.22)–(6.24). Starting with $u \in U$, following Fig. 6.1, the operator γ maps U onto $U^\#$. $A^\# = \gamma(u)$ is an input–output system form of the input u. This is Eq. (6.5). The operator ρ maps F to G. Equation (6.22) gives the auxiliary map to compute $G\left(A^\#\right) = B^\# \in Y^a$. At the top of Fig. 6.1, the generalized adjoint gives $F^a\left(B^\#\right) = B^\# F \in U^a$. This is Eq. (6.23). Finally, the equivalence operator, Eq. (6.24), gives $B^\# F \sim A^\#$. The remainder, and bulk of this chapter, is an analysis that provides boundedness and continuity conditions for the systems and transformations.

The upper portion of Fig. 6.1 shows the situation depicted in Fig. 1.1.

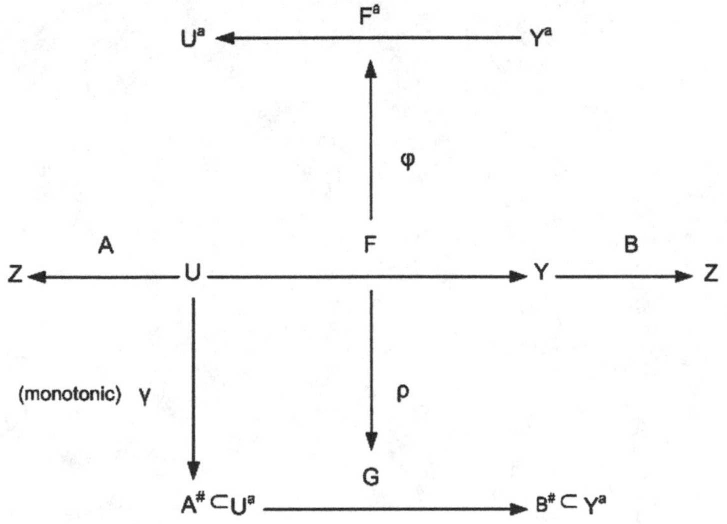

Fig. 6.1 Functional diagram

Chapter 7
Noise and Disturbance Bounds Using Adjoints

We consider the generalized adjoint in computing bounds for undesired inputs such as noise and disturbance. The method of adjoints has been discussed for traditional input–output systems; for example, in Laning and Battin [8]. Zarchan [22] gives a discussion of adjoints in connection with noise, disturbances, and other error sources in the homing loop of a missile autopilot. As discussed in these two references, the adjoint facilitates the calculation of noise and disturbance budgets and this will be reproduced, to some extent, in the context of generalized adjoints. We proceed by first separating the input into components, which we will view as various undesired inputs. Let the input be represented by

$$u = (u_1, u_2), \tag{7.1}$$

where the first component u_1 of u nominally represents the noise inputs and the second component u_2 nominally represents the disturbance inputs. The transfer function between noise input and the output and the disturbance input and the output are assumed to be different—which is usually the case. Toward the fulfillment of the hypotheses of Lemmas 24 through 31, we assume ζ is a constant $\overline{\zeta}$ in this chapter. We will need that γ be monotonic. Lemma 28 gives this if the input space (7.1) is a Hilbert space. The inner product for the compound input may be given by

$$\langle u, v\rangle = \langle (u_1, u_2), (v_1, v_2)\rangle = \langle u_1, u_2\rangle + \langle v_1, v_2\rangle \ . \tag{7.2}$$

With the tapered L_2 norm used on u_1 and u_2; and the inner product (7.2), the input space (7.1) is a Hilbert space. We represent the system in compound form as follows:

$$y = F(u) = F(u_1, u_2) \ . \tag{7.3}$$

D. Serakos, *Generalized Adjoint Systems*, SpringerBriefs in Optimization,
DOI 10.1007/978-3-319-16652-0_7

In this chapter, the $\mathcal{C}_N$ topology is used for F, although in the case of noise and disturbance inputs it may be more reasonable to use the sup-norm topology of bounded spaces, $\mathcal{C}_0$. Assume that the output response due to noise and disturbance inputs for an acceptable design is fixed and bounded. The objective is to determine maximum allowable noise and disturbance, i.e., determine a noise budget and a disturbance budget. Once these budgets have been determined, measures are taken so that excessive noise and disturbance do not enter the system. These measures may include filtering, design, and components. Typically the noise and disturbances enter the system independently. Therefore, the noise budget and disturbance budget are computed such that for any noise within the noise budget, together with any disturbance within the disturbance budget, the output response due to noise and disturbance is acceptable. We assume that the system is monotonic in the noise and disturbance, i.e., greater input noise/disturbance results in greater output noise/disturbance. Specifically, F monotonic in noise and disturbance means that given (u_1, u_2) and (u_1', u_2') such that $\|(u_1, u_2)\| \leq \|(u_1', u_2')\|$ implies $\|F(u_1, u_2)\| \leq \|F(u_1', u_2')\|$.

Using the standard method (not using adjoints), various u_1 and u_2 are fed into (7.3), together and separately, since F is nonlinear. By observing y, budgets for u_1 and u_2 would be determined. With adjoints, the budgets are simultaneously determined.

Consider the bound on the output due to the noise and disturbances. We need the output $\|y\| = \|F(u)\| \leq C$, where C is this bound. Define a subset of $\mathcal{B}$, $\mathcal{B}_C \triangleq \{B | B = \gamma(y), \|y\| = C\}$. Since we have that γ is monotonic function, we only need $B \in \mathcal{B}_C$ to correspond to $y's$ at the noise and disturbance bound, which is the edge of the shape in the transformed space. Define $\mathcal{A}_C \subset \mathcal{A} \triangleq \{A | A = F^a(B), B \in \mathcal{B}_C\}$. Going back to the input space, define $U \supset U_C \triangleq \gamma^{-1}(\mathcal{A}_C)$. Now compute

$$u_{1\,\max} = \max \|u_1\| \text{ such that } (u_1, u_2) = u \in U_C \ .$$

Similarly, define $u_{2\,\max}$. These are the noise and disturbance bounds. At this point we need the monotonicity of F mentioned above. This is analogous to portions of the situations presented in [8, 22].

In summary, bounds for noise and disturbance at the output are quantities we can set as design goals. Using the upper and left sides of Fig. 6.1, U is determined. The monotonicities of the system and the transformation γ are essential. The input end has the same monotonicity but does not have the previously mentioned edge to work with. Two cases may be addressed with regard to the system input. First, the case where the design is around a fixed input, or equivalently, an autonomous system. Second, take the input and disturbance together. The input u is normally modeled to enter the system at the same location as the disturbance d. This model would have $u_2 = u + d$. In this case, the disturbance bound discussed would be parcelled out between the input and the actual disturbance.

Chapter 8
Example

In this chapter, an illustrative example is presented. We illustrate some of the properties of the generalized adjoint system using a plant represented by a second degree integral operator.

Let $\mathcal{F}_t = \{(Y_0, F_t, U_0)\}$, be a space of causal input–output systems given by a second degree integral operator. The system equation is

$$F_t(u_0)(\sigma) = \int_0^\infty \int_0^\infty f(t+\sigma, \nu_1, \nu_2)\, u(\sigma-\nu_1)\, u(\sigma-\nu_2)\, d\nu_1 d\nu_2\,, \tag{8.1}$$

where $-\infty < \sigma \leq 0$.[1] The input space U_0 is the space of real-valued L_2 time functions over $(-\infty, 0]$ and Y_0 is the Banach space of real-valued L_2 time functions over $(-\infty, 0]$. The kernel f satisfies

$$f(t, v_1, v_2) = 0 \text{ for } v_1 < 0 \text{ or } v_2 < 0, \tag{8.2}$$

which is the usual causality condition. The weighted $\mathcal{C}_2(U_0, Y_0)$ norm will be used on F_t. If the kernel $f \in L_2\left(\Re^3\right)$, (with the weighting function) then F is bounded. Following calculations in Appendix B of [16] we get

$$\|y_0\|_0^2 = \int_{-\infty}^0 \|y(\sigma)\|^2\, w(-\sigma)\, d\sigma,$$

where w is a weighting function as in Eq. (2.1). Continuing,

$$= \int_{-\infty}^0 \left\| \int_0^\infty \int_0^\infty f(t+\sigma, \nu_1, \nu_2)\, u(\sigma-\nu_1)\, u(\sigma-\nu_2)\, d\nu_1 d\nu_2 \right\|^2 w(-\sigma)\, d\sigma$$

[1] This system is also used in Example 1 of [15].

D. Serakos, *Generalized Adjoint Systems*, SpringerBriefs in Optimization,
DOI 10.1007/978-3-319-16652-0_8

$$= \int_{-\infty}^{0} \left\| \int_{0}^{\infty} \int_{0}^{\infty} \frac{f(t+\sigma, \nu_1, \nu_2)}{(w(\nu_1-\sigma)\, w(\nu_2-\sigma))^{1/2}} u(\sigma-\nu_1)\, u(\sigma-\nu_2) \right.$$

$$\left. (w(\nu_1-\sigma)\, w(\nu_2-\sigma))^{1/2}\, d\nu_1 d\nu_2 \right\|^2 w(-\sigma)\, d\sigma \,.$$

Using the Schwarz inequality we have

$$\leq \int_{-\infty}^{0} \left[\int_{0}^{\infty} \int_{0}^{\infty} \frac{\| f(t+\sigma, \nu_1, \nu_2)\|^2}{w(\nu_1-\sigma)\, w(\nu_2-\sigma)} d\nu_1 d\nu_2 \right.$$

$$\left. \int_{0}^{\infty} \int_{0}^{\infty} \| u(\sigma-\nu_1)\, u(\sigma-\nu_2)\|^2 (w(\nu_1-\sigma)\, w(\nu_2-\sigma))\, d\nu_1 d\nu_2 \right] w(-\sigma)\, d\sigma$$

$$\leq \int_{-\infty}^{0} \int_{0}^{\infty} \int_{0}^{\infty} \frac{\| f(t+\sigma, \nu_1, \nu_2)\|^2}{w(\nu_1-\sigma)\, w(\nu_2-\sigma)} d\nu_1 d\nu_2 w(-\sigma)\, d\sigma \left(\|u_0\|^2 \right)^2 . \tag{8.3}$$

Then with $f \in L_2(\Re^3)$, define

$$\| f_t \|_w \triangleq \left(\int_{-\infty}^{0} \int_{0}^{\infty} \int_{0}^{\infty} \frac{\| f(t+\sigma, \nu_1, \nu_2)\|^2}{w(\nu_1-\sigma)\, w(\nu_2-\sigma)} w(-\sigma)\, d\nu_1 d\nu_2 d\sigma \right)^{1/2} < \infty \,. \tag{8.4}$$

Substituting (8.4) we get that (8.3) becomes

$$\| y_0 \|_0 \leq \| f_t \|_w \cdot \| u_0 \|_0^2 \,.$$

Hence, $\| F_t \|_0 \leq \| f_t \|_w$. Recall from (2.9) that $\| F \| = \sup_t \| F_t \|_0$; hence, if $\| f_t \|_w \leq$ Const. $< \infty$ we get that F is bounded as well. Consider the continuity of this system. We have

$$\| y_0 - y_0' \|_0^2$$

$$= \int_{-\infty}^{0} \| y(\sigma) - y'(\sigma) \|^2 w(-\sigma)\, d\sigma$$

$$= \int_{-\infty}^{0} \left\| \int_{0}^{\infty} \int_{0}^{\infty} f(t+\sigma, \nu_1, \nu_2)\, u(\sigma-\nu_1)\, u(\sigma-\nu_2)\, d\nu_1 d\nu_2 \right.$$

$$\left. - \int_{0}^{\infty} \int_{0}^{\infty} f(t+\sigma, \nu_1, \nu_2)\, u'(\sigma-\nu_1)\, u'(\sigma-\nu_2)\, d\nu_1 d\nu_2 \right\|^2 w(-\sigma)\, d\sigma$$

$$= \int_{-\infty}^{0} \left\| \int_{0}^{\infty} \int_{0}^{\infty} \frac{f(t+\sigma, \nu_1, \nu_2)}{(w(\nu_1-\sigma)\, w(\nu_2-\sigma))^{1/2}} (u(\sigma-\nu_1)\, u(\sigma-\nu_2) \right.$$

$$\left. - u'(\sigma-\nu_1)\, u'(\sigma-\nu_2))\, (w(\nu_1-\sigma)\, w(\nu_2-\sigma))^{1/2}\, d\nu_1 d\nu_2 \right\|^2 w(-\sigma)\, d\sigma$$

$$= \int_{-\infty}^{0} \left\| \int_{0}^{\infty} \int_{0}^{\infty} \frac{f(t+\sigma, \nu_1, \nu_2)}{(w(\nu_1-\sigma)\, w(\nu_2-\sigma))^{1/2}} \right.$$

$$\big(u(\sigma-\nu_1)\, u(\sigma-\nu_2) - u(\sigma-\nu_1)\, u'(\sigma-\nu_2)$$

$$+u(\sigma-\nu_1)\, u'(\sigma-\nu_2) - u'(\sigma-\nu_1)\, u'(\sigma-\nu_2)\big)$$

$$\left. (w(\nu_1-\sigma)\, w(\nu_2-\sigma))^{1/2}\, d\nu_1 d\nu_2 \right\|^2 w(-\sigma)\, d\sigma$$

using the Schwarz inequality,

$$\leq \int_{-\infty}^{0} \left[\int_{0}^{\infty} \int_{0}^{\infty} \frac{\| f(t+\sigma, \nu_1, \nu_2)\|^2}{w(\nu_1-\sigma)\, w(\nu_2-\sigma)} d\nu_1 d\nu_2 \right.$$

$$\int_{0}^{\infty} \int_{0}^{\infty} \big(u(\sigma-\nu_1)\, u(\sigma-\nu_2) - u(\sigma-\nu_1)\, u'(\sigma-\nu_2)$$

$$+u(\sigma-\nu_1)\, u'(\sigma-\nu_2) - u'(\sigma-\nu_1)\, u'(\sigma-\nu_2)\big)^2$$

$$\left. w(\nu_1-\sigma)\, w(\nu_2-\sigma)\, d\nu_1 d\nu_2 \right] w(-\sigma)\, d\sigma$$

$$= \int_{-\infty}^{0} \left[\int_{0}^{\infty} \int_{0}^{\infty} \frac{\| f(t+\sigma, \nu_1, \nu_2)\|^2}{w(\nu_1-\sigma)\, w(\nu_2-\sigma)} d\nu_1 d\nu_2 \right.$$

$$\int_{0}^{\infty} \int_{0}^{\infty} \big(u(\sigma-\nu_1)\big(u(\sigma-\nu_2) - u'(\sigma-\nu_2)\big)$$

$$+\big(u(\sigma-\nu_1) - u'(\sigma-\nu_1)\big)\, u'(\sigma-\nu_2)\big)^2$$

$$\left. w(\nu_1-\sigma)\, w(\nu_2-\sigma)\, d\nu_1 d\nu_2 \right] w(-\sigma)\, d\sigma$$

$$= \int_{-\infty}^{0} \left[\int_{0}^{\infty} \int_{0}^{\infty} \frac{\| f(t+\sigma, \nu_1, \nu_2)\|^2}{w(\nu_1-\sigma)\, w(\nu_2-\sigma)} d\nu_1 d\nu_2 \right.$$

$$\cdot 2 \cdot \int_{0}^{\infty} \int_{0}^{\infty} \Big(u^2(\sigma-\nu_1)\big(u(\sigma-\nu_2) - u'(\sigma-\nu_2)\big)^2$$

$$+\big(u(\sigma-\nu_1) - u'(\sigma-\nu_1)\big)^2 u'^2(\sigma-\nu_2)\Big)$$

$$\left. w(\nu_1-\sigma)\, w(\nu_2-\sigma)\, d\nu_1 d\nu_2 \right] w(-\sigma)\, d\sigma\, .$$

Hence,

$$\left\| y_0 - y_0' \right\|_0 \leq 2 \cdot \| f_t \|_w \cdot \max\left(\| u_0 \|_0, \left\| u_0' \right\|_0\right) \cdot \left\| u_0 - u_0' \right\|_0 . \tag{8.5}$$

Equation (8.5) gives that F is continuous. We get that F is equicontinuous for $\| f_t \|_w <$ Const. $< \infty$ and F uniformly continuous for U bounded.

Next, we consider the input space for the generalized adjoint system. Let $\mathcal{B}_t = \{(Y_t, F_t, U_0)\}$ be the space of causal linear time-varying input–output systems represented by

$$B_t(y_0)(\sigma) = \int_0^\infty b(t+\sigma, \nu)y(\sigma-\nu)d\nu, \tag{8.6}$$

where $-\infty < \sigma \leq 0$ and Z_0 the Banach space of real-valued L_2 time functions over $(-\infty, 0]$. Consider boundedness for $B_t \in \mathcal{B}_t$,

$$\begin{aligned}
\|z_0\|_0^2 &= \int_{-\infty}^0 \|z(\sigma)\|^2 w(-\sigma)\,d\sigma \\
&= \int_{-\infty}^0 \left\| \int_0^\infty b(t+\sigma,\nu)\,y(\sigma-\nu)\,d\nu \right\|^2 w(-\sigma)\,d\sigma \\
&= \int_{-\infty}^0 \left\| \int_0^\infty \frac{b(t+\sigma,\nu)}{w(\nu-\sigma)^{1/2}} y(\sigma-\nu)\,w(\nu-\sigma)^{1/2}\,d\nu \right\|^2 w(-\sigma)\,d\sigma\,.
\end{aligned}$$

Using the Schwarz inequality

$$\begin{aligned}
&\leq \int_{-\infty}^0 \left(\int_0^\infty \frac{\|b(t+\sigma,\nu)\|^2}{w(\nu-\sigma)} d\nu \int_0^\infty \|u(\sigma-\nu)\|^2 w(\nu-\sigma)\,d\nu \right) w(-\sigma)\,d\sigma \\
&\leq \int_{-\infty}^0 \int_0^\infty \frac{\|b(t+\sigma,\nu)\|^2}{w(\nu-\sigma)} w(-\sigma)\,d\nu d\sigma\ \|u_0\|_0^2\,.
\end{aligned}$$

Let

$$\|b_t\|_w \triangleq \left(\int_{-\infty}^0 \int_0^\infty \frac{\|b(t+\sigma,\nu)\|^2}{w(\nu-\sigma)} w(-\sigma)\,d\nu d\sigma \right)^{1/2} < \text{Const.} < \infty. \tag{8.7}$$

Under the condition (8.7), we have $\|B\| = \sup_t \|B_t\|_0 \leq \sup_t \|b_t\|_w$.

The generalized adjoint system $F_t^a : Y_0^a \to U_0^a$ of F_t is given by

$$\begin{aligned}
F_t^a(B_0)(u_0)(\sigma) &= (B_0 F_t)(u_0)(\sigma) = A_0(u_0)(\sigma) \\
&= \int_0^\infty b(\sigma,\nu) \int_0^\infty \int_0^\infty f(t+\sigma-\nu,\nu_1,\nu_2)u(\sigma-\nu-\nu_1) \\
&\quad u(\sigma-\nu-\nu_2)d\nu_1 d\nu_2 d\nu\,.
\end{aligned} \tag{8.8}$$

The input–output system A_0 in (8.8) is not in the form of an integral operator; however, it can be put in that form by interchanging the order of integration. By a change in variables (8.8) becomes

$$A_0(u_0)(\sigma) = \int_0^\infty b(\sigma,\nu) \int_{-\infty}^{\sigma-\nu} \int_{-\infty}^{\sigma-\nu} f(t+\sigma-\nu, \sigma-\nu-\nu_1, \sigma-\nu-\nu_2)$$
$$u(\nu_1)u(\nu_2)d\nu_1 d\nu_2 d\nu$$
$$= \int_{-\infty}^{\sigma} b(\sigma,\sigma-\nu) \int_{-\infty}^{\nu} \int_{-\infty}^{\nu} f(t+\nu, \nu-\nu_1, \nu-\nu_2)u(\nu_1)u(\nu_2)d\nu_1 d\nu_2 d\nu \, ,$$

and now interchanging the order of integration and using (8.2) we get

$$= \int_{-\infty}^{\sigma} \int_{-\infty}^{\sigma} \left(\int_{\max(\nu_1,\nu_2)}^{\sigma} b(\sigma,\sigma-\nu) f(t+\nu, \nu-\nu_1, \nu-\nu_2) d\nu \right)$$
$$u(\nu_1)u(\nu_2)d\nu_1 d\nu_2 \tag{8.9}$$
$$= \int_0^\infty \int_0^\infty \left(\int_{\max(\nu-\nu_1,\nu-\nu_2)}^{\sigma} b(\sigma,\sigma-\nu) f(t+\nu, \nu-\sigma+\nu_1, \nu-\sigma+\nu_2) d\nu \right)$$
$$u(\sigma-\nu_1)u(\sigma-\nu_2)d\nu_1 d\nu_2 \, , \tag{8.10}$$

which is in the form of a second order integral operator with the term inside the parentheses in (8.10) being the kernel. To consider conditions for the interchange in (8.9) define $\|b\|_{\infty,2} = \sup_\sigma \|b(\sigma,\cdot)\|_2$. Then, sufficient conditions to interchange the order of integration are $b \in L_{\infty,2}(\Re)$ (i.e., $\|b\|_{\infty,2} < \infty$), $f \in L_1(\Re^3)$ and $u \in L_2(\Re)$.

We verify various properties of the generalized adjoint which were presented earlier. To demonstrate that F^a is memoryless, as given by Proposition 11, consider two systems B^1, B^2 of the form (8.6) such that $\|B^1 - B^2\|_{s,0} = 0$, for $-\infty < s < 0$. This gives $b^1(t,\nu) = b^2(t,\nu)$ for $t \in (s,0)$ and $\nu \geq 0$ Then, using (8.8),

$$F_t^a(B_0^1 - B_0^2)(u_0)(\sigma) = (B_0^1 - B_0^2)F_t(u_0)(\sigma)$$
$$= \int_0^\infty (b^1(\sigma,\nu) - b^2(\sigma,\nu)) \int_0^\infty \int_0^\infty f(t+\sigma-\nu, \nu_1, \nu_2)$$
$$u(\sigma-\nu-\nu_1)u(\sigma-\nu-\nu_2)d\nu_1 d\nu_2 d\nu = 0 \, ,$$

where $s \leq \sigma \leq 0$, for all $u_0 \in U_0$. Therefore, $\|F_t^a(B_0^1 - B_0^2)\|_{s,0} = 0$, and F^a is memoryless (and causal).

To demonstrate the time invariance of F^a when F is time invariant, also in Proposition 11, let $f(t,\nu_1,\nu_2) = f(t',\nu_1,\nu_2)$ for all $t,t' \in \Re$. Consider (recalling R_s is the right translate from Definition 1 and R_s^* is the adjoint right translate from Chap. 3)

$$\left(\left(L_s^* F_t^a R_s^*\right) B_0\right)(u_0)(\sigma) = \left(L_s^* F_t^a (R_s B_0 L_s) u_0\right)(\sigma)$$
$$= (L_s (R_s B_0 L_s F_t) R_s u_0)(\sigma) = (B_0 L_s F_t R_s u_0)(\sigma) \, ,$$

where $\sigma < 0$. Now,

$$
\begin{aligned}
& L_s F_t\,(R_s u_0)\,(\sigma) \\
& = L_s \int_0^\infty \int_0^\infty f\,(t+\sigma, \nu_1, \nu_2)\, R_s u\,(\sigma - \nu_1)\, R_s u\,(\sigma - \nu_2)\, d\nu_1 d\nu_2 \\
& = L_s \int_0^\infty \int_0^\infty f\,(t+\sigma, \nu_1, \nu_2)\, u\,(\sigma - \nu_1 - s)\, u\,(\sigma - \nu_2 - s)\, d\nu_1 d\nu_2 \\
& = \int_0^\infty \int_0^\infty f\,(t+\sigma+s, \nu_1, \nu_2)\, u\,(\sigma + s - \nu_1 - s)\, u\,(\sigma + s - \nu_2 - s)\, d\nu_1 d\nu_2 \\
& = \int_0^\infty \int_0^\infty f\,(t+\sigma+s, \nu_1, \nu_2)\, u\,(\sigma - \nu_1)\, u\,(\sigma - \nu_2)\, d\nu_1 d\nu_2 \\
& = \int_0^\infty \int_0^\infty f\,(t+\sigma, \nu_1, \nu_2)\, u\,(\sigma - \nu_1)\, u\,(\sigma - \nu_2)\, d\nu_1 d\nu_2\,,
\end{aligned}
$$

by the time invariance of F. So

$$
\begin{aligned}
& \left(\left(L_s^* F_t^a R_s^*\right) B_0\right) (u_0)\,(\sigma) \\
& \quad = \int_0^\infty b\,(\sigma, \nu)\, L_s F_t\,(R_s u_0)\,(\sigma - \nu)\, d\nu \\
& \quad = \int_0^\infty b\,(\sigma, \nu) \int_0^\infty \int_0^\infty f\,(t + \sigma - \nu, \nu_1, \nu_2)\, u\,(\sigma - \nu - \nu_1) \\
& \quad \quad u\,(\sigma - \nu - \nu_2)\, d\nu_1 d\nu_2 d\nu \\
& \quad = B_0 F_t\,(u_0)\,(\sigma) = F_t^a (B_0)(u_0)(\sigma)\,,
\end{aligned}
$$

as in (8.8). Hence, F^a is time invariant.

Consider the form of (6.22) for the system (8.1) of this example,

$$
\begin{aligned}
& G(A^\#)(y)(t) = \left\| y_t - \tilde{F}_t(u_t^\#) \right\|_t \cdot \zeta(t) \\
& = \left(\int_{-\infty}^t \left\| y_t\,(\sigma) - \int_0^\infty \int_0^\infty f(\sigma, \nu_1, \nu_2) u^\#(\sigma - \nu_1) u^\#(\sigma - \nu_2) d\nu_1 d\nu_2 \right\|^2 d\sigma \right)^{1/2} \\
& \cdot \zeta(t)\,.
\end{aligned}
$$

The form of (6.23) and (6.24) for this example is (given f is one to one),

$$
\begin{aligned}
& (F^a(B^\#))u(t) = \left\| \tilde{F}_t(u_t) - \tilde{F}_t(u_t^\#) \right\|_t \cdot \zeta(t) \\
& = \left(\left\| \int_{-\infty}^t \int_0^\infty \int_0^\infty f(\sigma, \nu_1, \nu_2) \right.\right.
\end{aligned}
$$

$$\left(u(\sigma - \nu_1)u(\sigma - \nu_2) - u^{\#}(\sigma - \nu_1)u^{\#}(\sigma - \nu_2)\right) d\nu_1 d\nu_2 \Big\|^2 d\sigma\Big)^{1/2} \cdot \zeta(t)$$

$$\sim \left(\int_{-\infty}^{t} \left\|u(\sigma) - u^{\#}(\sigma)\right\|^2 d\sigma\right)^{1/2} \cdot \zeta(t)\,.$$

Note F is one to one if $f(t, v_1, v_2) \neq 0 \forall\, t, v_1 > 0, v_2 > 0$.

Chapter 9
Summary and Conclusions

In this book, we have defined a generalized adjoint system for a causal dynamical input–output system. The input–output system may be nonlinear and time varying. The generalized adjoint system is patterned after adjoints of bounded linear transformations in Banach space, and inherits various properties from the original input–output system. We have also shown that the generalized adjoint system can be used to represent the inverse of the original input–output system. Finally, the generalized adjoint system may be of some use in computing bounds for undesired inputs such as noise and disturbance to the original input–output system.

An input–output system is denoted by a triple (Y, F, U) where U is the input space, F is the system map, and Y is the output space. The generalized adjoint system of (Y, F, U) is the system (U^a, F^a, Y^a) where $Y^a = \{(Z, B, Y)\}$ and $U^a = \{(Z, A, U)\}$. The generalized adjoint system is defined by $A = F^a(B) = BF \in U^a$ for $B \in Y^a$. Note that the input and output spaces of the generalized adjoint system, Y^a and U^a, respectively, are themselves spaces of causal input–output systems. The causal input–output system $B \in Y^a$ may be viewed as a time function by considering its system trajectory: $t \to B_t$. The spaces Y^a and U^a are to be selected and specific choices were made to fulfill the needs of the book.

The results up to and including Proposition 18 concern the generalized adjoint system and the generalized adjoint map. These results include that the generalized adjoint system is always linear and memoryless. That if the original input–output system is time invariant and/or bounded, then the generalized adjoint is time invariant and/or bounded. The generalized adjoint map is the map from a family of input–output systems $\mathcal{F} = \{(Y, F^i, U)\}$ to their generalized adjoints $\mathcal{F}^a = \{(U^a, F^{ia}, Y^a)\}$ (same input and output spaces for the families). For bounded input–output systems, the generalized adjoint map is bounded and preserves translations. Under additional hypotheses, the generalized adjoint map is Lipschitz continuous and is bounded below, which implies its inverse is continuous. The Hahn–Banach theorem is used to obtain this result.

D. Serakos, *Generalized Adjoint Systems*, SpringerBriefs in Optimization,
DOI 10.1007/978-3-319-16652-0_9

In the first section of Chap. 6, Lemmas 19 through 22 show, in an algebraic sense, that the generalized adjoint system is one to one/onto when the original system is onto/one to one, and vice versa. In the second section of Chap. 6, a functional analysis of the generalized adjoint system representing the inverse of the original system is given. The input and output spaces of the generalized adjoint system are not the same as the output and input spaces of the original system. Because of this, an auxiliary input–output system (Y^a, G, U^a) is defined which is related to the original system and has the same input and output spaces as the generalized adjoint system, but switched. Define $A^\# \in U^a$ of the form $A^\#(u)(t) = \|u_t - u_t^\#\|_t \cdot \zeta(t)$, where ζ is an element of the equivalence class z, for some $z \in Z$, with $\zeta(t) \neq 0$ for all t, (the sup norm is used for Z). The set of all $A^\#$ is $\mathcal{A}^\# \subset U^a$. A map $\gamma : U \to \mathcal{A}^\#$ defined by $\gamma(u) = A^\#$. γ is one to one by construction. Lemmas 23 through 28 show the $A^\# \in \mathcal{A}^\#$ are causal, bounded, and uniformly continuous input–output systems; that γ is monotonic, bounded, and uniformly continuous; and that γ^{-1} is bounded and continuous. (Hence, γ is a homeomorphism.) The map G is defined by

$$G(A^\#)(y)(t) \triangleq \|(y - F(u^\#))_t\|_t \cdot \zeta(t) .$$

The map G is related to F in that if $y^\# = F(u^\#)$ and $A^\# \in \mathcal{A}^\#$ and $B^\# \in \mathcal{B}^\#$ are derived from $u^\#$ and $y^\#$, respectively, as described above (with γ), then $G(A^\#) = B^\#$. Let ρ be the map $\rho : \mathcal{F} \to \mathcal{G}$ ($\rho(F) = G$). Lemmas 29 through 31 show that the auxiliary systems G are memoryless, bounded, and continuous; that ρ is bounded and continuous; and that ρ^{-1} is bounded and continuous in a coarser topology. We showed F^a inverts G (hence F) by defining an equivalence relation for systems in $\mathcal{A}$, [see (6.23) and Fig. 6.1]. Lemmas 32 and 33 consider the boundedness and continuity of the equivalence relation.

Chapter 7 demonstrates how generalized adjoints may be used to compute bounds for undesired inputs such as noise and disturbance for an input–output system. This is similar to part of the adjoints of traditional input–output systems (see, for example, Laning and Battin [8]).

The results of Chaps. 4 and 5, and several of the results in Chap. 6 are illustrated in Chap. 8's example with a system represented by a second degree integral operator. The form of the auxiliary system is also shown with this type of system.

Appendix
On the Input–Output System Topology

This appendix presents various aspects of the input–output system topology that is used in this book. So that this appendix is self-contained, we include the definitions of fitted families (FFs) of time functions, and other relevant definitions, which are also found in the main body of this book.

Definition 34 ([11]). Let $\mathcal{L} = \mathcal{L}(\Re, E)$ be a linear space of time functions from $\Re$ into a Banach space E such that any translate of a function in $\mathcal{L}$ is also a function in $\mathcal{L}$. Let $\mathcal{N} = \{\|\cdot\|_{s,t}, -\infty < s < t < \infty\}$ be a family of seminorms on $\mathcal{L}$ satisfying the following conditions:

(1) For $f_1, f_2 \in \mathcal{L}$, if $f_1(\tau) = f_2(\tau)$ for $s < \tau \le t$, then $\|f_1 - f_2\|_{s,t} = 0$.
(2) Let L_τ denote shift to the left by τ. For all $f \in \mathcal{L}$, $\|\mathrm{L}_\tau f\|_{s-\tau,t-\tau} = \|f\|_{s,t}$.
(3) Let $r < s < t$. Then for all $f \in \mathcal{L}$, $\|f\|_{s,t} \le \|f\|_{r,t}$.
(4) Let $r < s < t$. Then for all $f \in \mathcal{L}$, $\|f\|_{r,t} \le \|f\|_{r,s} + \|f\|_{s,t}$.
(5) There exists $0 < \alpha \le \infty$ and $K \ge 1$ such that if $0 < t - r \le \alpha$ and $r < s < t$, then for all $f \in \mathcal{L}$, $\|f\|_{r,s} \le K\|f\|_{r,t}$.
The pair $(\mathcal{L}, \mathcal{N})$ is called an *FF of seminorms* on $\mathcal{L}$. The normed linear space formed from equivalence classes of functions in $\mathcal{L}$ with norm $\|\cdot\|_{s,t}$ is denoted $H_{s,t}$. The elements of $H_{s,t}$ are the equivalence classes determined by: $f \sim g$, $f, g \in \mathcal{L}$ if and only if $\|f - g\|_{s,t} = 0$. They are denoted $u_{s,t}$, $y_{s,t}$, etc. The set $\{H_{s,t}\}$, $-\infty < s < t < \infty$, is the *FF of normed linear spaces* given by $(\mathcal{L}, \mathcal{N})$.

For $f \in \mathcal{L}$, put $\|f\|^{s,t} \triangleq \sup_{s<\tau\le t} \|f\|_{s,\tau}$. An FF $(\mathcal{L}, \mathcal{N})$ and $\{H_{s,t}\}$, $-\infty < s < t < \infty$, can be augmented to include $\|\cdot\|_{-\infty,t}$ by taking the limit $s \to -\infty$, since by (3) of Definition 34 $\|f\|_{s,t}$ is monotone nondecreasing as $s \to -\infty$ with t fixed. Let $\mathcal{L}_0 = \{f \in \mathcal{L} \mid \lim_{s\to-\infty} \|f\|_{s,t} < \infty, t \in \Re\}$. For $f \in \mathcal{L}_0$, define

$$\|f\|_t \triangleq \lim_{s\to-\infty} \|f\|_{s,t} = \|f\|_{-\infty,t}\,. \tag{A.1}$$

D. Serakos, *Generalized Adjoint Systems*, SpringerBriefs in Optimization,
DOI 10.1007/978-3-319-16652-0

With the meaning of $(\mathcal{L}, \mathcal{N})$ thus extended, $\|\cdot\|_{s,t}$ is defined for $-\infty \leq s < t < \infty$. The *left-expanded FF of seminorms* is thereby defined and is denoted $(\mathcal{L}_0, \mathcal{N})$. It still satisfies all the Conditions (1), ..., (5).

We next define $\|\cdot\|_{s,\infty}$ and $H_{s,\infty}$. For an FF, this is done by taking the supremum. Let $\mathcal{L}_{00} = \{f \in \mathcal{L}_0 | \sup_t \|f\|_t < \infty\}$. For $f \in \mathcal{L}_{00}$, define

$$\|f\|_{s,\infty} \triangleq \sup_{t>s} \|f\|_{s,t} \; ; -\infty \leq s. \tag{A.2}$$

It may be readily verified that if $(\mathcal{L}, \mathcal{N})$ is an FF for indices satisfying $-\infty < s < t < \infty$ then, with definitions given by (A.1) and (A.2), $(\mathcal{L}_{00}, \mathcal{N})$ is an FF for indices satisfying $-\infty \leq s < t < \infty$ and satisfies Conditions (1)–(3), and (5) of Definition 34 for indices $-\infty \leq s < t \leq \infty$. $\{(\mathcal{L}_{00}, \mathcal{N}),\ \|\cdot\|_{s,t},\ -\infty \leq s < t \leq \infty\}$ is called the *expanded family of seminorms* determined by $(\mathcal{L}, \mathcal{N})$.

For $f \in \mathcal{L}_{00}$, we put

$$\|f\| \triangleq \sup_{t \in \Re} \|f\|_t \; = \|f\|_{-\infty,\infty} \,. \tag{A.3}$$

The normed linear space consisting of equivalence classes of functions in $\mathcal{L}_{00}$ with the norm (A.3) is called the *bounding space* H for the family $\{H_{s,t}\}$. For $-\infty \leq r < s < t < \infty$, and $g, h \in \mathcal{L}$; the *splice* of g and h over $(r, t]$ at s is defined and equals f if

$$f(\tau) = \begin{cases} g(\tau), \; r < \tau \leq s \\ h(\tau), \; s < \tau \leq t \end{cases}$$

belongs to $\mathcal{L}$. It is denoted $f_{r,t} = g_{r,s} \leftrightarrow h_{s,t}$. For $t = \infty$, the splice of g and h equals f if

$$f(\tau) = \begin{cases} g(\tau), \; r < \tau \leq s \\ h(\tau), \; s < \tau \end{cases}$$

belongs to $\mathcal{L}$. It is denoted $f_{r,\infty} = g^{r,s} \leftrightarrow h_{s,\infty}$.

Let Φ be a mapping from a normed linear space X into a normed linear space Y. As shown in Appendix A of [16], for any nonnegative integer N, the N-power norm for Φ is given by

$$[\![\Phi]\!]_{(N)} \triangleq \sup_{x \in X} \frac{\|\Phi(x)\|}{1 + \|x\|^N} \tag{A.4}$$

when the right side exists. The sub (N) is often omitted from the left-hand side of (A.4).

Lemma 35 ([16]). *Let M, N be positive integers, $N > M$. If $\|\Phi\|_{(M)} < \infty$ then $\|\Phi\|_{(N)} < \infty$.*

Proof. Assume $\|\Phi\|_{(M)} < \infty$. In the first place, let $\|x\| \geq 1$. Then

$$\|\Phi\|_{(N)} = \sup_{x \in X} \frac{\|\Phi(x)\|}{1 + \|x\|^N} \leq \sup_{x \in X} \frac{\|\Phi(x)\|}{1 + \|x\|^M} \leq \|\Phi\|_{(M)}.$$

In the second place, let $\|x\| < 1$. Then

$$\frac{\|\Phi(x)\|}{2} \leq \frac{\|\Phi(x)\|}{1 + \|x\|^M} \leq \frac{\|\Phi(x)\|}{1 + \|x\|^N} \leq \|\Phi(x)\|$$

or,

$$\frac{\|\Phi(x)\|}{1 + \|x\|^N} \leq \frac{2\,\|\Phi(x)\|}{1 + \|x\|^M}.$$

Hence, we conclude that

$$\|\Phi\|_{(N)} \leq 2 \cdot \|\Phi\|_{(M)}.$$ □

A mapping $F : U \to Y$ is called a *(global) input–output mapping* [or referred to as an input–output system (Y, F, U)]. The spaces U and Y are bounding spaces.

Definition 36. Let (Y, F, U) be an input–output system. F is a *causal mapping* and (Y, F, U) is a *causal system* if and only if for all t and for all $u, v \in U$ such that $\|u - v\|_t = 0$ it follows that $\|F(u) - F(v)\|_t = 0$.

If F satisfies Definition 36, it determines a mapping from U_t into Y_t, denoted $\tilde{F}_t$, that satisfies $\left\| \tilde{F}_t u_t - (Fu)_t \right\|_t = 0$. We call $\tilde{F}_t$ a *truncated input–output mapping* and define the *centered truncated input–output mapping* $F_t : U_0 \to Y_0$ by $F_t(u_0) \triangleq \mathrm{L}_t \tilde{F}_t \mathrm{R}_t(u_0)$.

Using (A.4) on the truncated system mapping, we have

$$\|F_t\|_{(N)} \triangleq \sup_{u_0} \frac{\|F_t(u_0)\|_0}{1 + \|u_0\|_0^N} = \sup_{u_t} \frac{\left\|\tilde{F}_t(u_t)\right\|_t}{1 + \|u_t\|_t^N} = \|\tilde{F}_t\|_{(N)}. \tag{A.5}$$

In [15, 20] the norm for the global input–output mapping is defined by

$$\|F\|^* \triangleq \sup_u \frac{\|F(u)\|}{1 + \|u\|^N}. \tag{A.6}$$

However, in this book input–output systems are themselves used as inputs to the generalized adjoint systems. Therefore, the norm for the global input–output mapping should be similar to the norm for an input, as in (A.3). In this book the global input–output mapping is defined by

$$\|F\| = \sup_t \|F_t\|_0 \triangleq \sup_t \|F_t\|_{-\infty,0}\,, \tag{A.7}$$

[16]. We can use subscripts on the operator norms similar to the subscripts in (A.3). The two global system norms (A.6) and (A.7) are related by

$$\|F\|^* \leq \|F\|.$$

Denote the causal bounded input–output mappings from U to Y by $\mathcal{D}_N(U,Y)$. The following lemma (a special case of Lemma A.1 in [16]) gives conditions so $\mathcal{D}_N(U,Y)$ is a Banach space.

Lemma 37 ([16]). *Let $Y_{-b,0}$ and Y be Banach spaces where $0 \leq b \leq \infty$. Then $\mathcal{D}_N(U_0, Y_{-b,0})$ is a Banach space with norm $\|\cdot\|_{-b,0}$ and $\mathcal{D}_N(U,Y)$ is a Banach space with norm $\|\cdot\|$.*

With (A.6) as the definition of boundedness, continuity does not imply boundedness for a linear system, which is usually the case. So, we define a stronger form of continuity.

Definition 38 ([16]). An input–output system (Y, F, U), $F \in \mathcal{D}_N(U,Y)$ is *compatibly continuous* if the truncated maps F_t are equicontinuous.

Our initial lemma relates continuity with compatible continuity.

Lemma 39 ([16]). *Compatible continuity is stronger than continuity.*

Proof. If $\|u - v\| = \sup_t \|u - v\|_t < \delta$, i.e., $\|u - v\|_t < \delta$ for all t, then $\left\|\tilde{F}_t(u_t) - \tilde{F}_t(v_t)\right\|_t < \varepsilon$, for all t. Hence, $\|F(u) - F(v)\| = \sup_t \left\|\tilde{F}_t(u_t) - \tilde{F}_t(v_t)\right\|_t < \varepsilon$. □

And now we have the following lemma:

Lemma 40 ([16]). *Let (Y, F, U) be a linear input–output system. Then compatible continuity is equivalent to boundedness.*

proof. (i) Consider boundedness: If (Y, F, U) is compatibly continuous, for all $\|u_0\|_0 = 1$ or $\|\delta u_0\|_0 = \delta$, $(\|F_t(\delta u_0)\|_0 < \varepsilon)$ then $\|F_t(u_0)\|_0 < \varepsilon/\delta$, for all t. Hence,

$$\frac{\|F_t(u_0)\|_0}{\|u_0\|_0} < \frac{\varepsilon}{\delta}\,,$$

for all t, and (Y, F, U) is bounded.

(ii) Consider compatible continuity: We first consider the observation that for F_t linear, the linear operator norm is equivalent to (A.5). Observe that

$$\|F_t\| = \sup_{u_0} \frac{\|F_t(u_0)\|_0}{1 + \|u_0\|_0^N} = \sup_{u_0} \frac{\|F_t(u_0)\|_0}{\|u_0\|_0} \cdot \frac{\|u_0\|_0}{1 + \|u_0\|_0^N}$$
$$\leq \sup_{u_0} \frac{\|F_t(u_0)\|_0}{\|u_0\|_0} = \|F_t\|_{(\text{linear})}.$$

Also,

$$\|F_t\| = \sup_{u_0} \frac{\|F_t(u_0)\|_0}{1 + \|u_0\|_0^N} \geq \sup_{\|u_0\|=1} \frac{\|F_t(u_0)\|_0}{1 + \|u_0\|_0^N} = \sup_{\|u_0\|=1} \frac{\|F_t(u_0)\|_0}{2}$$
$$= \frac{1}{2} \cdot \sup_{\|u_0\|=1} \|F_t(u_0)\|_0 = \frac{1}{2} \cdot \|F_t\|_{(\text{linear})}.$$

Putting these together, we have

$$\frac{1}{2} \cdot \|F_t\|_{(\text{linear})} \leq \|F_t\| \leq \|F_t\|_{(\text{linear})}, \tag{A.8}$$

which gives the equivalence. Getting to the compatible continuity, if (Y, F, U) is bounded,

$$\|F_t(u_0 - v_0)\|_0 \leq \|F_t\|_{(\text{linear})} \|u_0 - v_0\|_0$$
$$\leq 2 \cdot \|F_t\| \|u_0 - v_0\|_0 \leq 2 \cdot \|F\| \|u_0 - v_0\|_0 .$$

□

Denote the set of causal bounded, compatibly continuous input–output systems by $\mathcal{C}_N(U, Y)$. It may be shown that if $\alpha = +\infty$, $0 \in U$, and U is closed with respect to splicing, compatible continuity is equivalent to continuity [16].

Lemma 41 ([16]). *Let (Y, F, U) be a causal input–output system. If $\alpha = +\infty$, $0 \in U$ and U is closed with respect to splicing, compatible continuity is equivalent to continuity.*

Proof. Let $\|u_t - v_t\|_t < \delta$. Since $\alpha = +\infty$, for all $s < t$, $\|u_s - v_s\|_s < \mathcal{K} \cdot \delta$. This gives $\|u^t \leftrightarrow 0_{t,\infty} - v^t \leftrightarrow 0_{t,\infty}\| < \mathcal{K} \cdot \delta$. By the continuity of (Y, F, U), given $\varepsilon > 0$, we may choose $\mathcal{K} \cdot \delta$ (independent of t) such that $\|F(u^t \leftrightarrow 0_{t,\infty}) - F(v^t \leftrightarrow 0_{t,\infty})\| < \varepsilon$. This gives $\|F(u^t \leftrightarrow 0_{t,\infty}) - F(v^t \leftrightarrow 0_{t,\infty})\|_t < \varepsilon$, and by causality, $\left\|\tilde{F}_t(u_t) - \tilde{F}_t(v_t)\right\|_t < \varepsilon$. This is compatible continuity. □

The global input–output system being compatibly continuous gives that the truncated systems are continuous. This may not be said if the global input–output system is continuous, as demonstrated by the following example:

Example 42 ([16]). In this example a linear input–output system (Y, F, U) is given such that the global map is bounded under (A.6) and the truncated map is unbounded. Let U be the space of real-valued Lebesgue measurable functions with the exponentially tapered fitted family of seminorms

$$\|u\|_t = \int_{-\infty}^{t} |u(\tau)| \, e^{a(t-\tau)} d\tau. \tag{A.9}$$

Let Y be the same as U but with

$$\|y\|_t = \int_{-\infty}^{t} |y(\tau)| \, e^{b(t-\tau)} d\tau \tag{A.10}$$

such that $-\infty < a < b < 0$. Let F be the identity. Since F is linear, the ones in the denominators of (A.5) and (A.6) may be omitted, see (A.8). Consider the class of inputs

$$\bar{u}_T(t) = \begin{cases} e^{xt}, & -T < t \leq 0 \\ = 0 \text{ elsewhere} \end{cases}$$

such that $a < x < b$. We have that

$$\llbracket F_0 \rrbracket \geq \frac{\|F_0(\bar{u}_T)\|_0}{\|\bar{u}_T\|_0} = \frac{\left| \int_{-T}^{0} e^{(x-b)\tau} d\tau \right|}{\left| \int_{-T}^{0} e^{(x-a)\tau} d\tau \right|}$$

$$= \frac{\left| \frac{1}{(x-b)} \right|}{\left| \frac{1}{(x-a)} \right|} \cdot \frac{\left| e^{(x-b)\tau} \big]_{-T}^{0} \right|}{\left| e^{(x-a)\tau} \big]_{-T}^{0} \right|} = \left(\frac{x-a}{x-b} \right) \left| \frac{1 - e^{(b-x)T}}{1 - e^{(a-x)T}} \right| .$$

Consider the right factor, since $(a - x)$ is negative, the denominator tends towards one as T increases. Since $(b - x)$ is positive, the numerator becomes larger and larger without bound, as T increases. Hence, F_0 is not bounded. Next, consider the global map. For all $u \in U$, define

$$m \triangleq \sup_{p} \int_{p}^{p+1} |u(\tau)| \, d\tau < \infty , \tag{A.11}$$

for all $p \in \Re$. Taking (A.9) and using (A.11), we obtain

$$\|u\| = \sup_{t} \int_{-\infty}^{t} |u(\tau)| \, e^{a(t-\tau)} d\tau \geq m \cdot e^{a} .$$

Similarly taking (A.10)

$$\|y\| = \sup_t \int_{-\infty}^{t} |y(\tau)|\, e^{b(t-\tau)} d\tau$$
$$= \sup_t \sum_{k=0}^{\infty} \int_{t-(k+1)}^{t-k} |y(\tau)|\, e^{b(t-\tau)} d\tau \le m \cdot \sum_{k=0}^{\infty} e^{bk}.$$

Substituting in (A.6), we see that

$$\|F\|^{*} \le \frac{m \cdot \sum_{k=0}^{\infty} e^{bk}}{m \cdot e^{a}} = \frac{\sum_{k=0}^{\infty} e^{bk}}{e^{a}} = C < \infty.$$

Since C is independent of u, F is bounded using the $\|\cdot\|^{*}$ norm, (A.6).

It is the author's opinion that it is unsettling when a causal system may have a bounded global map while the induced truncated systems are unbounded. This example illustrates this possibility when the global norm is given by (A.6). These types of systems, in the author's opinion, should be excluded or treated specially. When the global norm is given by (A.7), this situation is automatically excluded.

References

1. Campobasso, M.S., Duta, M.C., Giles, M.B.: Adjoint methods for turbomachinery design. In: International Symposium On Air Breathing Engines (2001)
2. Deimling, K.: Nonlinear Functional Analysis. Springer, New York (1985)
3. Desoer, C.A., Vidyasagar, M.: Feedback Systems: Input-Output Properties. Academic, New York (1975)
4. Dunford, N., Schwartz, J.T.: Linear Operators, Part I: General Theory. Interscience Publishers Inc., New York (1957)
5. Freudenberg, J.S., Middleton, R.H., Solo, V.: Stabilization and Disturbance Attenuation Over a Gaussian Communication Channel. IEEE Trans. Autom. Control, **55**(3), 795–799 (2010)
6. Heitman, G.K.: Identifiability of classes of input-output systems. J. Math. Anal. Appl. **188**, 774–797 (1994)
7. Kailath, T.: Linear Systems, Prentice-Hall Inc., New Jersey (1980)
8. Laning, J.H. Jr, Battin, R.H.: Random Processes in Automatic Control. McGraw-Hill, New York (1956)
9. Morency, C., Luo, Y., Tromp, J.: Finite-frequency kernels for wave propagation in porous media based upon adjoint methods. Geophys. J. Int. **179**, 1148–1168 (2009)
10. Pires, C., Miranda, P.M.A.: Tsunami waveform inversion by adjoint methods. J. Geophys. Res. **106**(C9), 19773–19796 (2001)
11. Root, W.L.: Considerations regarding input and output spaces for time-varying systems. Appl. Math. Optim. **4**, 365–384 (1978)
12. Root, W.L.: A note on state trajectories of causal input-output systems. In: Proceedings Twenty-First Annual Allerton Conference on Communications, Control and Computing, The University of Illinois at Urbana-Champaign, Allerton House, Monticello, IL (1983)
13. Root, W.L., Serakos, D.: On causal input-output systems. In: Proceedings Twenty-Third Annual Allerton Conference on Communications, Control and Computing, The University of Illinois at Urbana-Champaign, Allerton House, Monticello, IL (1985)
14. Root, W.L., Serakos, D.: Generalized adjoint systems. In: Proceedings 20th Princeton Conference on Information Sciences and Systems, Department of Electrical Engineering, Princeton University (1986)
15. Root, W.L., Serakos, D.: The state of dynamical input-output systems as an operator. J. Math. Anal. Appl. **225**, 224–248 (1998)
16. Serakos, D.: Topics in input-output systems theory: feedback systems with tapered input spaces, state and generalized adjoint systems. Ph.D. Dissertation, The University of Michigan (1988)

D. Serakos, *Generalized Adjoint Systems*, SpringerBriefs in Optimization,
DOI 10.1007/978-3-319-16652-0

17. Serakos, D.: Stability in feedback systems with tapered and other special input spaces. IEEE Trans. Autom. Control **37**(8), 1256–1260 (1992)
18. Serakos, D.: Stability, Aim Bias compensation and noise sensitivity of Phalanx CIWS control system. In:Proceedings of IEEE Regional Conference on Aerospace Control Systems. IEEE Control Systems Society, Westlake Village, CA, USA 25–27 (1993)
19. Serakos, D.: Generalized adjoint systems inverses. In: Proceedings 35th Annual Allerton Conference on Communication, Control and Computing, Department of Electrical Engineering, The University of Illinois, Urbana-Champaign (1997)
20. Serakos, D.: State space consistency and differentiability conditions for a class of causal dynamical input-output systems. NSWCDD/TR-08/3, Dahlgren Division Naval Surface Warfare Center, Dahlgren, Virginia 22448-5100 (2008)
21. Serakos, D.: State Space Consistency and Differentiability. Springer, New York (2015)
22. Zarchan, P.: Tactical and strategic missile guidance. Prog. Astronaut. Aeronaut. **124**, AIAA (1990)